Objective Soil Science & Agricultural Chemistry

For Class and Competitive Examinations

Objective Soil Science & Agricultural Chemistry

For Class and Competitive Examinations

D K SARKAR

Former Professor

Acharya N G Ranga Agricultural University

Andhra Pradesh

Former Associate Dean, S K Y College of Agricultural Sciences, Andhra Pradesh

BSP **BS Publications**

An Imprint of **BSP Books Pvt. Ltd.**

4-4-309/316, Giriraj Lane,

Sultan Bazar, Hyderabad - 500 095

Objective Soil Science & Agricultural Chemistry

by **D K Sarkar**

Published by

BSP **BS Publications**

An Imprint of **BSP Books Pvt. Ltd.**

4-4-309/316, Giriraj Lane, Sultan Bazar,
Hyderabad - 500 09
Phone: 040 - 23445688
e-mail: info@bspbooks.net
www.bspbooks.net

ISBN: 978-93-49562-00-4 (Hardback)

Preface

Sometime back, after the announcement of the results of a competitive examination, few students of mine lamented that they could not secure desired ranks despite their hard work and painstaking preparation. Their remarks blended with despair stirred my sense, eventually leading to my logical thinking that for a mediocre student, the required inputs for cracking competitive examinations are qualitatively far different from routine class examinations. Primarily it is so because the books and reading materials, methods of instructions and patterns of evaluation in pursuit of a degree are ideally designed for buildup of concept and comprehension of a subject and do not fittingly serve the purpose of securing high ranks in today's competitive examinations. Logic, format and modus operandi of today's competitive examinations are strikingly different. To cite an instance, one has to select the right answer from several closely matching incorrect ones. The sticky wicket situation of several closely related wrong choices, often with cadence amidst the right one, adds an element of confusion to an average student in picking up the correct answer. And so, things often go haywire. A low score and poor rank in an examination where candidates far outnumber the slots is the result. To overcome, an average student has to rigorously practice answering such questions, so that the deficiency is won over and the student no longer gets confused when things matter in an examination hall with limited time to work out. For that to be so, students need examination-oriented tailor-made books. There is particular need for such books in the subject of Soil Science and Agricultural Chemistry. This specially designed book is an attempt in that direction.

The book is no substitute for textbooks and is specially crafted only for acquiring the required punch for a high score in examinations on the subject. In more lucid terms, the questions set in the book are to be answered only with the comprehension and knowledge gathered from reading text books.

I believe that the book shall be of much use for the purpose for which it has been conceived and articulated.

D. K. SARKAR

Contents

I. Choosing the Right Answer Option

Underline the most appropriate answer

1. Tourmaline is a
   (a) primary mineral (b) secondary mineral
   (c) clay mineral (d) carbonate mineral

2. The objective of photointerpretation in imagery is
   (a) identification of the objects
   (b) recognition of the objects
   (c) judging the significance of the objects
   (d) all the three

3. Transporting agency involved in the formation of deltas are
   (a) gravity (b) wind
   (c) ice (d) streams and rivers

4. The diagnostic subsurface horizon (endopedon) that is light coloured and poor in silicate clays and oxides of iron and aluminum is
   (a) Umbric (b) Spodic (c) Albic (d) Sombric

5. In remote sensing,
   (a) electromagnetic radiation interacts with the target
   (b) electromagnetic radiation is emitted from the target
   (c) both the two
   (d) neither of the two

6. Clay particles have their diameter
   (a) less than 0.002 mm (b) 0.02 to 0.002 mm
   (c) less than 0.005 mm (d) 0.05 to 0.002 mm

7. The pF of soil water at field capacity is
   (a) 0.33 (b) 2.53 (c) 4.1 (d) 4.5

8. Amount of water required to irrigate a land of 0.5 ha upto a depth of 5 cm is
   (a) 250 L (b) 2500 L (c) 25000 L (d) 250000 L

9. The mineral most resistant to chemical weathering is
   (a) feldspar (b) mica (c) quartz (d) olivine

10. The graduations of the axes of the triangular textural diagram used for assigning the textural class of soils are in increments of
    (a) 5 per cent (b) 7.5 per cent
    (c) 10 per cent (d) 12.5 per cent

11. The cation associated with the flocculation of soil colloids is
    (a) calcium (b) sodium (c) potassium (d) hydrogen

12. The availability of this micronutrient increases with increase in soil pH
    (a) iron (b) copper
    (c) manganese (d) molybdenum

13. Average optimum C:N ratio of cultivated soils is
    (a) 5:1 (b) 10:1 (c) 100:1 (d) 200:1

14. The pF of soil water at hygroscopic point is
    (a) 1 (b) 2.53 (c) 4.1 (d) 4.53

15. The bulb of a soil thermometer is kept molded at an angle of
    (a) 60^0 (b) 90^0 (c) 120^0 (d) 180^0

16. The optical properties of a water body depend upon
    (a) absorption of radiation by the dissolved substances
    (b) absorption of radiation by the particulate matter in suspension
    (c) scattering of radiation by the particulate matter in suspension
    (d) all the three

17. Eutrophication refers to the pollution load of
    (a) soil (b) water bodies
    (c) atmosphere (d) lithosphere

18. The Cation Exchange Capacity of this clay mineral is the highest
    (a) kaolinite (b) vermiculite
    (c) montmorillonite (d) chlorite

19. The internal surface of this clay mineral is the highest
    (a) halloysite (b) illite
    (c) montmorillonite (d) chlorite

20. This mineral is a 2:1:1 type layer lattice clay mineral
    (a) smectite (b) illite
    (c) montmorillonite (d) chlorite

21. Amount of Ca^{2+} ions required for cation exchange with 2 mg of H^+ ions is
    (a) 2 mg (b) 4 mg (c) 40 mg (d) 80 mg

22. Particle density of common mineral soils is in the range of
    (a) 2.60 to 2.76 Mg m^{-3} (b) 2.75 to 3.5 Mg m^{-3}
    (c) 1.30 to 1.38 Mg m^{-3} (d) 1.38 to 1.76 Mg m^{-3}

23. Tourmaline is a secondary mineral containing
    (a) molybdenum (b) boron
    (c) manganese (d) chlorine

24. This insecticide is nonpersistent in soil
    (a) DDT (b) malathion (c) BHC (d) aldrin

25. Application of this fungicide mitigates copper deficiency in soils
    (a) Blitox (b) Carbendazim
    (c) Dithane M45 (d) Dithane Z-78

26. Cathode rays are deflected by
    (a) electric field
    (b) magnetic field
    (c) both electric field and magnetic field
    (d) neither electric field nor magnetic field

27. This crop among the four is the most erosion resistant crop
    (a) rice (b) maize
    (c) soybean (d) green gram

28. The fertilizer that contains 20.6% nitrogen is
    (a) urea (b) ammonium sulphate
    (c) ammonium nitrate (d) calcium ammonium nitrate

29. Toxic substance dhurrin produced in sorghum is a/an
    (a) nonprotein amino acid (b) alkaloid
    (c) cyanogenic glycoside (d) phenolic substance

30. The ionic form of nitrogen preferred by rice in the early stages of crop growth is
    (a) ammonium (b) nitrate
    (c) nitrite (d) amide

31. The crop that ushered in Green Revolution in India in late seventies and early eighties is
    (a) rice (b) wheat (c) maize (d) millets

32. *Kharif* and *rabi* apart, the other crop season in India
    (a) Varsati (b) Sharadi
    (c) Zaid (d) Mousami

33. Release of mineral potassium is a major source of the nutrient in soils rich in
    (a) kaolinite (b) illite
    (c) montmorillonite (d) chlorite

34. With increase in soil aeration done through tillage, loss of gaseous nitrogen through denitrification
    (a) increases (b) decreases
    (c) remains unaltered (d) may increase or decrease

35. Actinomycetes are the most active microorganisms responsible for the decomposition of organic matter in the following stage
    (a) initial stage (b) mid stage
    (c) final stage (d) all stages

36. The bottom most category of soil classification in USDA Soil Taxonomy is
    (a) Family (b) Series
    (c) Subgroup (d) Suborder

37. The clay mineral that helps formation of cracks in black soils upon drying is
    (a) kaolinite (b) illite
    (c) montmorillonite (d) vermiculite

38. A phosphorus containing mineral is
    (a) apatite (b) sylvinite
    (c) kainite (d) brucite

39. Soils suitable for plantation crops are
    (a) laterite soils (b) black soils
    (c) alluvial soils (d) sandy soils

40. A brucite sheet is a structural component of the layer lattice structure of the clay mineral
    (a) chlorite (b) chlorapatite
    (c) montmorillonite (d) illite

41. Olsen's method is used for the determination of available phosphorus in soils that are
    (a) strongly acidic (b) acidic
    (c) saline (d) neutral to alkaline

42. The most abundant of all the minerals present in igneous rocks is
    (a) feldspars (b) micas
    (c) quartz (d) pyroxenes

43. Per cent CO_2 content of soil air is around
    (a) 0.04% (b) 0.15 (c) 0.4 (d) 1.45

44. Bedrock native to the soil is absent in
    (a) alluvial soils (b) black soils
    (c) red soils (d) forest soils

45. *Kari* is the local name for
    (a) alluvial soils (b) acid sulphate soils
    (c) red soils (d) black soils

46. An important chemical reaction taking place in puddled paddy soils is
    (a) conversion of $FeCl_2$ to $FeCl_3$
    (b) conversion of $FeCl_2$ to Fe_2S_3
    (c) conversion of $FeCl_3$ to Fe_2O_3
    (d) conversion of $FeCl_3$ to $FeCl_2$

47. A potassium containing mineral is
    (a) apatite (b) sylvanite (c) albite (d) sylvinite

48. According to FCO Specifications for Biofertilizers in India, the stipulated level for contamination is 'no contamination' at the following level of dilution
    (a) 10^2 (b) 10^3 (c) 10^5 (d) 10^7

49. Soils best suited for dryland agriculture are
    (a) laterite soils (b) black soils
    (c) alluvial soils (d) red soils

50. The bulk density and particle density of a soil are 1.4 and 2.8 g/cm^3, respectively. Now, if the porosity of the soil is increased to 75%, the soil's bulk density is changed as follows
    (a) decreased to its half (b) increased by its half
    (c) doubled (d) increased by its one fourth

51. Black frost in plants occurs due to such temperature and Relative Humidity (RH)
    (a) temperature $< 0^0$ C and low RH
    (b) temperature $> 0^0$ C and low RH
    (c) temperature $> 0^0$ C and high RH
    (d) temperature $< 0^0$ C and high RH

52. Wavelength range of electromagnetic radiation apt for the process of photosynthesis in plants is
    (a) 400 to 760 nm (b) 760 to 1500 nm
    (c) 100 to 200 nm (d) 200 to 400 nm

53. The most abundant of all the minerals in the earth's crust are
(a) feldspars (b) micas
(c) quartz (d) pyroxenes

54. In Soil Taxonomy, a possible Order for the alluvial soils is
(a) Ultisol (b) Alfisol
(c) Entisol (d) Histosol

55. Saline soils are to be of common occurrence in
(a) low rainfall areas (b) high rainfall areas
(c) low temperature soils (d) high altitude areas

56. Acid soils are to be of common occurrence in
(a) low rainfall areas
(b) high rainfall areas
(c) low temperature regions
(d) soils with high depth water table

57. Hydraulic conductivity of soils follows the following order
(a) Ca-soil > Na-soil > K- soil (b) Ca-soil < Na-soil < K- soil
(c) Ca-soil > K-soil > Na- soil (d) K-soil > Ca-soil > Na- soil

58. Regur is the local name for
(a) alluvial soils (b) acid sulphate soils
(c) red soils (d) black soils

59. Chalka soil is the local name for
(a) black soils (b) acid sulphate soils
(c) red sandy loam soils (d) brown forest soils

60. Expanding type clay minerals are dominantly present in
(a) black soils (b) red soils
(c) lateritic soils (d) alluvial soils

61. Swelling shrinkage property is a prominent characteristic of
(a) red soils (b) black soils
(c) lateritic soils (d) alluvial soils

62. Amount of $CaCO_3$ required for the neutralization of 4 milli equivalents of H^+ is
(a) 4 mg (b) 8 mg (c) 200 mg (d) 400 mg

63. Saline soils are characterized by these characteristics
    (a) pH < 8.5, EC > 4 dS/m at 25^{0}C, ESP < 15
    (b) pH > 8.5, EC > 4 dS/m at 25^{0}C, ESP > 15
    (c) pH > 8.5, EC < 4 dS/m at 25^{0}C, ESP >15
    (d) pH > 8.5, EC > 4 dS/m at 25^{0}C, ESP < 15

64. Chemical transformation of hematite to limonite involves the following chemical process
    (a) oxidation (b) reduction
    (c) hydrolysis (d) hydration

65. The exchangeable sodium content of a soil is 34.5 mg/100 g soil. Assuming that CEC of the soil is 8 cmol (p^+)/kg soil, pH is > 8.5 and electrical conductivity is < 4 dS/m at 25^{0}C, the soil in question is
    (a) acid soil (b) saline soil
    (c) alkali soil (d) saline alkali soil

66. The cation associated with the deflocculation and dispersion of soil colloids is
    (a) calcium (b) sodium
    (c) potassium (d) hydrogen

67. Young and relatively immature soils belong to the following soil Order
    (a) Alfisol (b) Mollisol
    (c) Entisol (d) Vertisol

68. Addition of this substance during collection and storage of farm yard manure helps decreasing the loss of nitrogen in gaseous form
    (a) ammonium sulphate (b) limestone powder
    (c) gypsum (d) powdered basic slag

69. These igneous rocks are formed at considerable depth below the earth's surface
    (a) extrusive rocks (b) plutonic rocks
    (c) stratified rocks (d) clastic rocks

70. A mineral that can be described by its perfect cleavage is
    (a) hematite (b) quartz
    (c) feldspar (d) micas

71. At field capacity, moisture is held to the soil at a potential of
    (a) 1 atmosphere (b) $\frac{1}{3}$ atmosphere
    (c) 15 atmospheres (d) 31 atmospheres

72. According to FCO Specifications of Fertilizers, biuret content of fertilizer grade urea, per cent by weight, maximum, is
(a) 1.0 (b) 1.5 (c) 2.0 (d) 0.5

73. The plant nutrient that is recycled in soil through atmosphere is
(a) nitrogen (b) phosphorus
(c) potassium (d) calcium

74. In soil, *Nitrosomonas* bacteria carry out the process of oxidation of
(a) organic-N to amide-N (b) organic-N to NH_4^+
(c) NH_4^+ to NO_2^- (d) NO_2^- to NO_3^-

75. If in the ribbon behaviour test with a moist bolus, there is no coherence (ball formation) and single grains adhere to fingers, then the soil texture is
(a) sand (b) loamy sand
(c) sandy loam (d) sandy clay loam

76. ICAR-National Bureau of Soil Survey and Land Use Planning is located at
(a) New Delhi (b) Dehradun (c) Bhopal (d) Nagpur

77. Two of the common factors that promote soil erosion are
(a) contour bunding and overgrazing
(b) contour bunding and mixed farming
(c) shifting cultivation and overgrazing
(d) contour bunding and shifting cultivation

78. The mobility of ammonium in soils with respect to that of nitrate is
(a) lower (b) higher
(c) same (d) lower or higher

79. Weathering of basaltic lava produces
(a) alluvial soils
(b) black soils
(c) coarse textured sandy soils
(d) soils poor in exchangeable bases

80. Soils suitable for growing cashew nut are
(a) black soils (b) red laterite soils
(c) peat soils (c) alluvial soils

81. 'Chroma' which defines the purity of soil colour does have a scale of
    (a) 0 to 5 (b) 0 to 10 (c) 0 to 20 (d) 0 to 100

82. Karewas, a unique soil/soil deposit is found in the state of
    (a) Assam (b) Meghalaya
    (c) Jammu and Kashmir (d) Chhattisgarh

83. The fertilizer that contains ammonium and nitrate nitrogen in 1:1 proportion is
    (a) ammonium sulphate nitrate
    (b) ammonium sulphate
    (c) ammonium phosphate
    (d) calcium ammonium nitrate

84. Toxic cyanogenic glycoside dhurrin is produced in
    (a) maize (b) ragi (c) bajra (d) sorghum

85. International Institute of Rice Research is located in
    (a) New Delhi (b) Cuttack
    (c) Philippines (d) Mexico City

86. Whiptail disease in cauliflower is caused due to the deficiency of
    (a) manganese (b) molybdenum
    (c) boron (d) copper

87. Swelling shrinkage property is a prominent characteristic of the soils belonging to the Order
    (a) Alfisol (b) Ultisol (c) Spodosol (d) Vertisol

88. Soils with good physical properties are
    (a) acid soils (b) alkali soils
    (c) saline soils (d) calcareous soils

89. The factor that is of no significance in determining the scale in aerial photography
    (a) focal length (b) flying height
    (c) ground elevation (d) none of the three

90. When an electromagnetic radiation falls upon a transparent medium,
    (a) angle of incidence = angle of reflection
    (b) angle of incidence = angle of refraction
    (c) angle of incidence = sum of angle of reflection and angle of refraction
    (d) angle of refraction = sum of angle of incidence and angle of reflection

91. The metallic cofactor associated with the enzyme urease is
(a) cobalt (b) copper (c) nickel (d) manganese

92. The process of conversion of simple inorganic plant available forms of nutrients to their complex organic plant unavailable forms is called
(a) mineralization (b) immobilization
(c) assimilation (d) aminization

93. Potassium in its ionic form is fixed in the layer lattice structure of this clay mineral
(a) kaolinite (b) montmorillonite
(c) illite (d) chlorite

94. A farmer for his ½ ha field had to apply urea @ 250 kg/ha, but had to opt for CAN due to nonavailability of urea. The amount of CAN required is (N contents of CAN and urea are 25 and 46, respectively)
(a) 115 kg (b) 230 kg (c) 345 kg (d) 460 kg

95. P to P_2O_5 conversion factor is
(a) 1.21 (b) 0.83 (c) 2.29 (d) 0.44

96. According to Stokes' law, the terminal velocity of a spherical particle falling in a viscous medium is inversely proportional to the
(a) particle density of the falling particle
(b) viscosity of the medium
(c) radius of the falling particle
(d) acceleration due to gravity

97. According to FCO Specifications for the standards for Biofertilizers, the total viable count in terms of colony forming units (CFU) per gram of powder, granules or carrier material of *Rhizobium* must be a minimum of
(a) 5×10^3 cells (b) 5×10^5 cells
(c) 5×10^7 cells (d) 5×10^{10} cells

98. Soil consistence limits with varying levels of water in soil were proposed by
(a) Marburt (b) Munsell (c) USDA (d) Atterberg

99. Bulk density of a soil increases with increase in
(a) porosity (b) compaction
(c) organic matter content (d) tillage

100. One millimicron is equal to
(a) 10^{-2} m (b) 10^{-3} m (c) 10^{-6} m (d) 10^{-9} m

101. A field crop requires nitrogen @ 200 kg ha^{-1} from two sources viz. neem cake (5% N) and urea (46% N), in the ratio of 1:3. The amounts of neem cake and urea to be applied to a field of 0.5 ha are
   (a) 200 kg neem cake and 100 kg urea
   (b) 300 kg neem cake and 140 kg urea
   (c) 500 kg neem cake and 163 kg urea
   (d) 500 kg neem cake and 250 kg urea

102. Shifting cultivation practiced in North East India is called
   (a) Tehri cultivation (b) Sawami cultivation
   (c) Jhoom cultivation (d) Harhi cultivation

103. The size of a micro watershed is
   (a) 100 to 1000 ha (b) 1000 to 5000 ha
   (c) 5000 to 1000 ha (d) > 5000 ha

104. World Soil Day is celebrated on
   (a) 5th June (b) 5th September
   (c) 14th October (d) 5th December

105. The crop among the four that does have the minimum water requirement is
   (a) maize (b) black gram (c) sugarcane (d) cotton

106. The method used for the determination of available phosphorus in acid soils is
   (a) Olsen's method
   (b) Bray's 1 method
   (c) DTPA method
   (d) ammonium acetate method

107. The plant nutrient that imparts resistance to insect pests and diseases is
   (a) phosphorus (b) potassium
   (c) calcium (d) magnesium

108. If the electrical conductivity of a soil is 1.5 dS m^{-1}, the soluble salt concentration of the soil may prove
   (a) critical only for germination
   (b) sensitive only for the growth of salt sensitive crops
   (c) likely to cause salt injury to many crops
   (d) likely to cause salt injury to most crops

109. River transported delta soils of Mahanadi, Krishna and Godavari rivers are rich in
(a) black soils (b) red soils
(c) laterites (d) alluvium

110. According to the USDA Soil Classification system, the Order for the red soils is
(a) Mollisol (b) Vertisol (c) Aridisol (d) Alfisol

111. Soils of the Order Vertisol that swell on wetting and shrink on drying are suitable for
(a) barley (b) mustard
(c) rye (d) plantation crops

112. All other factors remaining equal, the power of the divalent cations to replace the cations on the soil exchange complexes is in the following order
(a) $Ca^{2+} < Mg^{2+} < Ba^{2+} < Sr^{2+}$ (b) $Mg^{2+} < Ca^{2+} < Sr^{2+} < Ba^{2+}$
(c) $Ba^{2+} < Sr^{2+} < Ca^{2+} < Mg^{2+}$ (d) $Ba^{2+} < Mg^{2+} < Ca^{2+} < Sr^{2+}$

113. Direct use of finely ground rock phosphate is recommended as a phosphatic fertilizer in soils that are
(a) neutral (b) strongly acidic
(c) alkaline (d) strongly alkaline

114. Chromatography was discovered by
(a) Mikhail Tswett (b) Kirchhoff
(c) Walsh and Alkermade (d) Frumkin

115. Stratification is often observed in the formation of this class of rocks
(a) igneous (b) sedimentary
(c) metamorphic (d) fire

116. The reference salt used in assigning the Salt Index of fertilizers for assessing the hazard of salt build up following their application in soils is
(a) sodium nitrate (b) potassium nitrate
(c) ammonium nitrate (d) calcium nitrate

117. In Soil Taxonomy, the Order for a soil under the Subgroup Thaptic Cryaquand is
(a) Aridisol (b) Andisol (c) Gelisol (d) Alfisol

118. Nitrogen fixing organism Cyanobacteria were earlier known as
(a) *Rhizobium* (b) *Azotobacter*
(c) Blue green algae (d) *Azolla*

119. Diagnostic surface horizon (epipedon) that is primarily sandy is
(a) Mollic (b) Umbric (c) Grossarenic (d) Histic

120. Oxygen Diffusion Rate (ODR) of soils is measured by
(a) Neutron Scattering method
(b) Pressure Plate Apparatus method
(c) Robinson pipette method
(d) Platinum microelectrode method

121. The seventeenth entrant in the list of essential plant nutrients is
(a) cobalt (b) nickel (c) silicon (d) chlorine

122. The oxidizing agent used in the determination of available nitrogen content of soils by Subbaiah and Asija's method is
(a) potassium dichromate (b) potassium permanganate
(c) sodium hydroxide (d) hydrogen peroxide

123. The growth of plant roots ceases when ODR of the soil ($g/cm^2/min$) drops to about (multiplied by 10^{-8})
(a) 200 (b) 100 (c) 50 (d) 20

124. Humus is
(a) crystalline in nature (b) amorphous in nature
(c) inorganic soil colloid (d) layer lattice soil colloid

125. Substitution of Si^{4+} ion by Al^{3+} ion in silicate clays gives rise to
(a) temporary charge (b) permanent charge
(c) pH dependent charge (d) variable charge

126. Ion exchange process is
(a) temperature independent (b) takes place on mass basis
(c) reversible (d) irreversible

127. PBS of soils is expressed in terms of
(a) m.e./100 g soil (b) cmol (p^+)/kg soil
(c) cmol (e^-)/kg soil (d) it is a number with no unit

128. Soil microorganisms next only to bacteria in abundance are
(a) fungi (b) algae
(c) actinomycetes (d) protozoa

129. Diagnostic subsurface horizon (endopedon) that is thin and slowly permeable, and does have a reddish brown to black coloured iron or manganese pan is
(a) Gypsic (b) Calcic (c) Placic (d) Folistic

130. This micronutrient is a component of the enzyme nitrogenase
    (a) manganese (b) molybdenum
    (c) boron (d) copper

131. The number of Orders in USDA system of Soil Classification is
    (a) eight (b) ten (c) twelve (d) fourteen

132. Ploughing of the land in hilly areas cutting across the slope is called
    (a) terracing (b) contour ploughing
    (c) ridge ploughing (d) bunding

133. The spectral region of the electromagnetic spectrum which passes through the atmosphere without much attenuation is known as
    (a) ozone window (b) ozone hole
    (c) black hole (d) atmospheric window

134. Diagnostic subsurface horizon (endopedon) that is intensely weathered, rich in sesquioxides and 1:1 type layer lattice clays is
    (a) Umbric (b) Spodic (c) Oxic (d) Sombric

135. Orthoclase through the process of carbonation is transformed to the mineral
    (a) kaolinite (b) microcline (c) albite (d) leucite

136. In agricultural chemistry, the term 'tankage' refers to a manure obtained from
    (a) carcasses of animals
    (b) residues of vegetable oil extracting plants
    (c) night soil
    (d) domestic urban wastes

137. Manufacture of ammonia by Haber Bosch process through the union of elemental nitrogen and hydrogen in 1:3 mole ratio is a process of the following kind
    (a) exothermic (b) endothermic
    (c) exergonic (d) spontaneous

138. Two of the common factors that deter soil erosion are
    (a) contour bunding and overgrazing
    (b) contour bunding and mixed farming
    (c) shifting cultivation and overgrazing
    (d) contour bunding and shifting cultivation

139. The microorganisms that are more important for the decomposition of lignin and other complex organic molecules and so play a more significant role in the later stages of decomposition of organic matter are
(a) bacteria (b) fungi
(c) actinomycetes (d) protozoa

140. In compact soils with no porosity, the process of nitrification stops at the following step
(a) organic N $\rightarrow NH_4^+$ (b) $NH_4^+ \rightarrow NO$
(c) $NH_4^+ \rightarrow NO_2^-$ (d) $NO_2^- \rightarrow NO_3^-$

141. One of the followings processes is not involved in chemical weathering of rocks and minerals
(a) hydration (b) hydrolysis
(c) dehydration (d) hydrogenation

142. Some species of this genus of bacteria are leaf-residing nitrogen fixers in some plants
(a) *Methylobacterium* (b) *Azotobacter*
(c) *Frankia* (d) *Beijerinckia*

143. N^{13}, N^{14} and N^{15} are atomic forms of the element nitrogen containing unequal number of
(a) electrons (b) neutrons
(c) protons (d) all the three

144. This microbe is a nonsymbiotic nitrogen fixer in soil
(a) *Rhizobium* (b) *Azotobacter*
(c) *Frankia* (d) *Anabaena*

145. The nutrient made available to plants by Vesicular Arbuscular Mycorrhizae is
(a) nitrogen (b) phosphorus
(c) potassium (d) sulphur

146. Organic materials in water bodies like lakes pollute the water due to
(a) eutrophication
(b) rise in BOD
(c) growth of natural population of aquatic bacteria
(d) all the three

147. One of the followings is not a direct agent of physical weathering of rocks and minerals
(a) water (b) wind (c) vegetation (d) glaciers

148. The process of denitrification takes place more favourably in
(a) submerged soils
(b) compact soils
(c) puddled soil
(d) all the three

149. By definition, heavy metals are those metals which are heavier than water by
(a) 2.9 times (b) 5 times (c) 10 times (d) 13.6 times

150. The total surface area of the clay minerals kaolinite, illite, montmorillonite and chlorite is in the following order
(a) montmorillonite < chlorite < kaolinite < illite
(b) chlorite < illite < montmorillonite < kaolinite
(c) chlorite < montmorillonite < illite < kaolinite
(d) kaolinite < illite < chlorite < montmorillonite

151. With an increase in soil p^H, the available pool of the micronutrient cations
(a) decreases (b) increases
(c) may increase or decrease (d) remains unaffected

152. The mobility of nitrate in soil system with respect to that of ammonium is
(a) lower (b) higher
(c) same (d) may be lower or higher

153. Diagnostic subsurface horizon (endopedon) that is characterized by accumulation of insoluble limestone and dolomite is
(a) Gypsic (b) Salic (c) Natric (d) Calcic

154. The wavelength range of the entire electromagnetic spectrum is
(a) 10^{-10} to 10^{8} m (b) 10^{-15} to 10^{8} m
(c) 10^{-9} to 10^{6} m (d) 10^{-8} to 10^{8} m

155. A radio isotope of phosphorus (atomic number 15) of much use in soil research is
(a) P^{29} (b) P^{30} (c) P^{31} (d) P^{32}

156. Soil moisture potential expressed in terms of bars at field capacity is
(a) 0.33 (b) 1 (c) -15 (d) -31

157. Advanced stage of sheet erosion is
   (a) Splash erosion (b) Gully erosion
   (c) Rill erosion (d) Tunnel erosion

158. According to the USDA Soil Classification system, a possible Order for the peat and muck soils is
   (a) Spodosol (b) Histosol (c) Oxisol (d) Gelisol

159. The analytical instrument used in assay for working with radioactive isotopes such as P^{32} is
   (a) atomic emission spectrophotometer
   (b) flame emission spectrophotometer
   (c) mass spectrometer
   (d) Geiger Muller Counter

160. Peat soils are generally
   (a) black, heavy and neutral (b) black, heavy and acidic
   (c) black, light and saline (d) black, light and alkaline

161. The crop that requires the highest Oxygen Diffusion Rate is
   (a) rice (b) wheat (c) potato (d) pea

162. The oxidizing agent used in the determination of organic carbon content of soils by Walkley and Black's Wet Oxidation method is
   (a) potassium dichromate (b) potassium permanganate
   (c) sodium thiosulphate (d) hydrogen peroxide

163. Diagnostic subsurface horizon (endopedon) that is characterized by accumulation of organic matter and oxides of iron and aluminium is
   (a) Kandic (b) Spodic (c) Argillic (d) Sombric

164. Specific heat of a body is defined as the ratio of its heat capacity to its
   (a) volume (b) mass
   (c) density (d) thermal conductivity

165. The chemical added for the inclusion of nitrate nitrogen in the determination of total nitrogen content of soils by Kjeldahl method is
   (a) sulphuric acid (b) salicylic acid
   (c) sodium nitrite (d) Devarda's alloy

166. The number of soil textural classes in the triangular textural diagram is
   (a) 8 (b) 10 (c) 12 (d) 14

167. The unit for expressing the electrical conductivity of soils at a specified temperature is
   (a) dS/m (b) mmhos/cm
   (c) both the two (d) neither of the two

168. The exchangeable Ca^{2+} content of a soil is 120 mg/100 g soil. Assuming that CEC of the soil is 8 cmol (p^+)/kg soil, per cent Ca^{2+} saturation of the soil is
(a) 15 (b) 37.5 (c) 75 (d) 80

169. The two active factors of soil formation are
(a) climate and parent material
(b) climate and biosphere
(c) climate and relief
(d) biosphere and parent material

170. Soil scientist who categorized the factors of soil formation into active and passive factors is
(a) Joffe (b) Jenny (c) Marbut (d) Dokuchaev

171. In soil, *Nitrobacter* bacteria carry out the following oxidation process
(a) organic-N to amide-N (b) organic-N to NH_4^+
(c) NH_4^+ to NO_2^- (d) NO_2^- to NO_3^-

172. The plant nutrient that has a rapid absorption rate in foliar spray is
(a) potassium (b) magnesium
(c) sulphur (d) molybdenum

173. A crop that has very high salt tolerance is
(a) cotton (b) sugar beet
(c) barley (d) rapeseed

174. The Salinity Class of irrigation water that has TDS level of 0.5 to 1.5 g/L and is rated as 'Doubtful for use for crops' is
(a) C1 (b) C2 (c) C3 (d) C4

175. In India, the highest land area under cultivation is comprised by
(a) alluvial soils (b) black soils
(c) laterite soils (d) red soils

176. The solum of a soil is comprised of
(a) A horizon (b) A and B horizons
(c) A, B and C horizons (d) A, B, C and D horizons

177. Microorganisms that are efficient producers of antibiotics are
(a) bacteria (b) fungi
(c) actinomycetes (d) protozoa

178. In Soil Taxonomy, the soils of the cold tundra region that possess permafrost are classified under the Order
(a) Histosol (b) Andisol (c) Spodosol (d) Gelisol

179. In the fractionation of humus, the constituent of the latter that remains insoluble in both acid and alkali is
(a) fulvic acid (b) humic acid
(c) polyuronic acid (d) humin

180. At hygroscopic coefficient, soil water is held with a force of
(a) $\frac{1}{3}$ atmosphere (b) 1 atmosphere
(c) 15 atmospheres (d) 31 atmospheres

181. Regolith of a soil is comprised by the following horizon/s
(a) A and B horizons
(b) C horizon
(c) B and C horizons
(d) A, B and C horizons

182. The soil that is not suitable for use in nursery is
(a) alluvial (b) black (c) red (d) laterite

183. In alkaline pH, the ionic form in which phosphorus exists is
(a) $H_2PO_4^-$ (b) HPO_4^{2-} (c) PO_4^{3-} (d) $P_2O_7^{4-}$

184. 'Die back' physiological disease in citrus is caused due to the deficiency of
(a) boron (b) copper
(c) zinc (d) molybdenum

185. A concentrated organic manure is
(a) FYM (b) compost
(c) vermicompost (d) neem cake

186. In Soil Taxonomy, a prominent characteristic of the area such as eroded, degraded etc. is included in naming a
(a) Subgroup (b) Family
(c) Series (d) Phase

187. The fertile soils of the Gangetic plains locally known as *Bhangar* and *Khaddar* are
(a) alluvial soils (b) red soils
(c) laterite soils (d) lateritic soils

188. The study of the different aspects of rocks such as their formation, composition, structure and properties is called
(a) geology (b) metrology
(c) petrology (d) mineralogy

189. Monocalcium phosphate is the phosphorus containing component of the phosphatic fertilizer
    (a) single super phosphate
    (b) triple super phosphate
    (c) both single and triple super phosphates
    (d) neither single nor triple super phosphate

190. Specific heat (c) of a substance, say soil, is expressed as
    (a) $c = \frac{\text{heat capacity}}{\text{mass}}$
    (b) $c = \frac{\text{heat capacity}}{\text{heat flux}}$
    (c) $c = \frac{\text{heat capacity}}{\text{volume}}$
    (d) $c = \frac{\text{heat capacity}}{\text{temperature gradient}}$

191. $CaSO_4$ is a constituent of the phosphatic fertilizer
    (a) single super phosphate
    (b) triple super phosphate
    (c) urea ammonium phosphate
    (d) diammonium phosphate

192. Conversion of complex organic nitrogen to simple inorganic nitrogen like ammonium compounds as done in the case of mineralization is a process of
    (a) carbonation
    (b) oxidation
    (c) reduction
    (d) hydration

193. With increase in soil pH, the anion exchange capacity of soils
    (a) decreases
    (b) increases
    (c) may increase or decrease
    (d) no relationship between the two

194. Fly ash increasingly being used as an amendment in agriculture is a byproduct of
    (a) iron and steel plants
    (b) cement manufacturing units
    (c) coal based thermal power plants
    (d) textile industry

195. ICAR-Indian Institute of Soil Science is located in
    (a) New Delhi
    (b) Karnal
    (c) Bhopal
    (d) Hyderabad

196. 'Value' which defines the relative brightness of soil colour does have a scale of
    (a) 0 to 5
    (b) 0 to 10
    (c) 0 to 20
    (d) 0 to 100

197. The concept of pF scale for expressing soil moisture tension was proposed by
    (a) Jenny (b) Schofield
    (c) Sorenson (d) Baver

198. Diagnostic subsurface horizon (endopedon) that is characterized by accumulation of silicate clays is
    (a) Kandic (b) Spodic (c) Argillic (d) Agric

199. On analysis, a soil with near neutral pH recorded an electrical conductivity of 5 dS/m at 25^0 C, cation exchange capacity of 18 m.e./100 g soil and exchangeable sodium of 45 mg/100 g soil. The soil in question is
    (a) alkali soil (b) sodic soil
    (c) saline alkali soil (d) saline soil

200. The enzyme concerned with Biological Nitrogen Fixation is
    (a) nitrogenase (b) nitrogen fixase
    (c) nitrate reductase (d) nitrite reductase

201. The term Ustipsamment signifies the following hierarchical category in Soil Taxonomy
    (a) Suborder (b) Great Group
    (c) Subgroup (d) Family

202. Terracing is an effective method of soil conservation in
    (a) Plain area (b) Hilly area
    (c) Desert area (d) High rainfall area

203. The gas produced during the anaerobic decomposition of organic matter in flooded paddy fields is
    (a) methane (b) ammonia
    (c) sulphur dioxide (d) carbon dioxide

204. Oxygen Diffusion Rate (O.D.R.) in soils is expressed in terms of
    (a) μg Oxygen cm^{-2} min^{-1} (b) ml Oxygen cm^{-2} min^{-1}
    (c) μg Oxygen m^{-1} min^{-1} (d) ml Oxygen m^{-1} min^{-1}

205. The vast mountain range of north-western part of India India running from Gujarat to Delhi through Rajasthan and Haryana that that acts a barrier for the clouds to shift eastwards to the lower Himalayas, thus contributing to the climate of north india is
    (a) Satpura (b) Sivalik (c) Vindhya (d) Aravalli

206. In the fractionation of humus, the constituent of the latter that is soluble in both alkali and acid is
(a) fulvic acid (b) humic acid
(c) polyuronic acid (d) lignin

207. A typical example of a fossil fuel is
(a) petroleum (b) liquid hydrogen
(c) biodiesel (d) ethanol

208. At pH 4, phosphorus exists in the following ionic form
(a) $H_2PO_4^-$ (b) HPO_4^{2-} (c) PO_4^{3-} (d) $P_2O_7^{4-}$

209. 'Little leaf' in cotton is caused due to the deficiency of
(a) copper (b) manganese
(c) zinc (d) molybdenum

210. If the particle density and per cent pore space of a soil are 2.6 g/cm^3 and 45 respectively, the bulk density of the soil is
(a) 1.37 g/cm^3 (b) 1.43 g/cm^3
(c) 1.55 g/cm^3 (d) 1.76 g/cm^3

211. *Akiochi* disease in rice is caused due to the excess formation of this gas in waterlogged rice fields
(a) methane (b) nitrous oxide
(c) ammonia (d) hydrogen sulphide

212. A solution of single super phosphate tests
(a) acidic (b) neutral
(c) slightly alkaline (d) highly alkaline

213. Basic slag is a byproduct of this industry
(a) cement (b) iron and steel
(c) refractory (d) textiles

214. According to FCO specifications, the minimum per cent sulphur content (*w/w*) of the fertilizer grade ammonium sulphate is
(a) 16 (b) 20.6 (c) 23.0 (d) 46.0

215. Single super phosphate is manufactured by acidulation of rock phosphate with
(a) sulphuric acid (b) hydrochloric acid
(c) nitric acid (d) phosphoric acid

216. Plants absorb phosphorus preferentially in the form of
(a) $H_2PO_4^-$ (b) HPO_4^{2-}
(c) both $H_2PO_4^-$ and HPO_4^{2-} (d) PO_4^{3-}

217. At wilting point, soil water is held with a force of
(a) $\frac{1}{3}$ atmosphere (b) 1 atmosphere
(c) 15 atmospheres (d) 31 atmospheres

218. If the bulk density and per cent pore space of a soil are 1.43 g/cm^3 and 45 respectively, the particle density of the soil is
(a) 1.37 g/cm^3 (b) 1.50 g/cm^3
(c) 2.60 g/cm^3 (d) 2.76 g/cm^3

219. Metallic element that is a constituent of chlorophyll is
(a) calcium (b) magnesium
(c) manganese (d) copper

220. Triple super phosphate is manufactured by acidulation of rock phosphate with
(a) sulphuric acid (b) hydrochloric acid
(c) nitric acid (d) phosphoric acid

221. Buttoning in cauliflower is caused due to the deficiency of
(a) nitrogen (b) phosphorus
(c) potassium (d) sulphur

222. Basic slag, a byproduct of iron and steel industry and a source of phosphorus, can be used for the reclamation of
(a) saline soils (b) alkali soils
(c) saline alkali soils (d) acid soils

223. The horizon that is not a part of the solum of a soil
(a) A horizon (b) B horizon
(c) C horizon (d) B and C horizons

224. These soils have high buffering capacity
(a) sandy soils (b) sandy loam soils
(c) sandy clay loam soils (d) clay soils

225. P_2O_5 content of nitrophosphates fertilizers is
(a) entirely water soluble
(b) entirely water insoluble
(c) entirely citrate insoluble
(d) partly water soluble and partly citrate soluble

226. India does not manufacture this fertilizer
(a) ammonium sulphate (b) nitrophosphates
(c) muriate of potash (d) NPK complexes

227. Removal of top soil by running water with the formation of small channels that can later be removed by normal tillage is
   (a) sheet erosion (b) gully erosion
   (c) rill erosion (d) splash erosion

228. Soil forming process in which soil materials like silica, sesquioxides etc. washed down from the upper soil depths by percolating water are deposited in the lower layers is called
   (a) eluviation (b) illuviation
   (c) podzolization (d) humification

229. In the Wind Erosion Equation, E= f (I, K, C, L, V), the notation I denotes
   (a) soil erodibility factor (b) surface roughness factor
   (c) climatic factor (d) field length factor

230. In the fractionation of humus, the constituent of the latter that is soluble in alkali, but insoluble in acid is
   (a) fulvic acid (b) humic acid
   (c) polyuronic acid (d) humin

231. A part of the negative charge in humus arises from the dissociation of this functional group of the organic compounds present in the colloid
   (a) aldehyde (b) ketone
   (c) amino (d) enolic OH

232. The wavelength of the following electromagnetic radiations follows the order
   (a) infrared > visible > X-ray > ultraviolet
   (b) visible > infrared > ultraviolet > X-ray
   (c) X-ray > visible > infrared > ultraviolet
   (d) infrared > visible > ultraviolet > X-ray

233. The buffering capacity of a soil increases with increase in the content of
   (a) silt (b) clay
   (c) organic matter (d) both clay and organic matter

234. The pF of soil water at wilting point is
   (a) 1 (b) 2.53 (c) 4.19 (d) 4.5

235. If the bulk density and particle density of a soil are 1.43 and 2.60 g/cm^3 respectively, the per cent pore space of the soil is
   (a) 25 (b) 40 (c) 45 (d) 55

236. Brown heart in turnip is caused due to the deficiency of
   (a) nitrogen (b) sulphur
   (c) boron (d) molybdenum

237. Basic slag, a byproduct of iron and steel industry, is a source of of this plant nutrient
   (a) phosphorus (b) potassium
   (c) nitrogen (d) manganese

238. Soil forming process in which soil materials like silica, sesquioxides etc. are washed down from the upper soil depths by percolating water is called
   (a) eluviation (b) illuviation
   (c) podzolization (d) humification

239. Podzolization soil formation process is favoured under
   (a) cool humid climate (b) high rainfall
   (c) acidic parent material (d) all the three

240. The SAR of an irrigation water is 7.5. The irrigation water is rated as
   (a) S1 Class. Low sodium hazard
   (b) S2 Class. Medium sodium hazard
   (c) S3 Class. High sodium hazard
   (d) S4 Class. Very high sodium hazard

241. A part of the negative charge in humus arises from the dissociation of this functional group of the organic compounds present in the colloid
   (a) aldehyde (b) ketone
   (c) amino (d) phenolic OH

242. The buffering capacity of a soil increases with increase in
   (a) bulk density (b) particle density
   (c) porosity (d) cation exchange capacity

243. Lime requirement of an acid soil increases with increase in the content of
   (a) silt (b) clay
   (c) organic matter (d) both clay and organic matter

244. Soil structure best suited for crop growth is
   (a) columnar (b) crumb
   (c) platy (d) prismatic

245. Soil forming process in which oxides of iron and aluminium washed down by percolating water from the upper soil depths are deposited in the lower layers is called
   (a) desalinization (b) gleization
   (c) podzolization (d) laterization

246. In the Wind Erosion Equation, E = f (I, K, C, L, V), the notation I denoting soil erodibility factor does have the dimension of

(a) mass (b) mass/unit time

(c) mass/unit area (d) mass/unit area/unit time

247. Savanna ecoregion refers to a biome of

(a) tropical rain forest

(b) mixed woodland-grassland

(c) large desert tract

(d) dense forest under temperate climate

248. Irrigation method best suited for undulating topography is

(a) flooding (b) basin method

(c) furrow irrigation (d) sprinkler method

249. Available water in the soil ranges between

(a) field capacity and wilting point

(b) field capacity and hygroscopic coefficient

(c) saturation point and hygroscopic coefficient

(d) wilting point and hygroscopic coefficient

250. In mycorrhizae, the relationship between the root cells of higher plants and the fungi is

(a) symbiotic (b) commensalism

(c) parasitism (d) nonsymbiotic

251. The role of *Bacillus megatherium*, a microbial biofertilizer, is to

(a) fix soluble phosphorus

(b) precipitate soluble phosphorus

(c) solubilize insoluble phosphorus

(d) immobilize soluble phosphorus

252. The transporting agency involved in the formation of alluvial soils is

(a) gravity (b) water (c) wind (d) ice

253. Gypsum process is widely used for the manufacture of this fertilizer

(a) ammonium sulphate

(b) calcium ammonium nitrate

(c) single super phosphate

(d) triple super phosphate

254. On analysis, a soil was reported to contain 0.065% nitrogen. Considering the average C:N of agricultural soils to be 10:1 to 15:1, the per cent organic matter content of the soil is
(a) 0.650 to 0.975 (b) 1.121 to 1.681
(c) 1.300 to 1.950 (d) 1.300 to 2.200

255. If the pOH of a solution is 4, H^+ ion concentration of the solution in mole/litre is
(a) 10^{-4} (b) 10^{-14} (c) 10^{-10} (d) 10^{4}

256. In India, rock phosphate deposits are found in this area
(a) Bundelkhand in Uttar Pradesh
(b) Mayurbhanj in Odisha
(c) Mussoorie in Uttarakhand
(d) Terai in West Bengal

257. Manufacture of muriate of potash either from sylvite or sylvinite does not require any strong acid or heat treatment because both the potassium minerals are
(a) water soluble (b) water insoluble
(c) occur at a shallow depth (d) are free from impurities

258. Soil forming process in which silica is washed down by percolating water from the upper soil depths and then deposited in the lower layers resulting in the accumulation of iron and oxides in the upper layers is called
(a) desalinization (b) gleization
(c) podzolization (d) laterization

259. The SAR of an irrigation water source is 28.5. The irrigation water is rated as
(a) S1 Class. Low sodium hazard
(b) S2 Class. Medium sodium hazard
(c) S3 Class. High sodium hazard
(d) S4 Class. Very high sodium hazard

260. Large part of the negative charge in humus arises from the dissociation of this functional group of the organic compounds present in the colloid
(a) aldehyde (b) ketone
(c) carboxyl (d) amino

261. In the Wind Erosion Equation, E= f (I, K, C, L, V), the notation K denotes
(a) soil erodibility factor (b) ridge roughness factor
(c) climatic factor (d) field length factor

262. The transporting agency in the formation of loess deposits of parent materials or soils is
(a) gravity (b) water (c) wind (d) ice

263. Denitrifying bacteria are
(a) obligate anaerobes (b) facultative anaerobes
(c) obligate aerobes (d) facultative aerobes

264. The fertilizer that has the highest per cent nitrogen content is
(a) urea solution (b) nitrogen solution
(c) aqua ammonia (d) anhydrous ammonia

265. Bangalore method of composting was developed by
(a) Hutchison and Richards (b) Dr C. N. Acharya
(c) Howard and Ward (d) Fowler and Ridge

266. The catalyst used in the manufacture of ammonia from N_2-H_2 (1:3) mixture at 550^0 C is
(a) nickel (b) mercury
(c) activated iron oxide (d) Devarda's alloy

267. Flotation process is widely used for the manufacture of this fertilizer
(a) ammonium sulphate (b) single super phosphate
(c) muriate of potash (d) urea

268. Den process is widely used for the manufacture of this fertilizer
(a) triple super phosphate (b) single super phosphate
(c) nitrophosphates (d) diammonium phosphate

269. Sulphate of Potash is manufactured by
(a) Langbeinite process (b) Manheim furnace process
(c) Both the two (d) Neither of the two

270. A mix of ammonium chloride and ammonium hydroxide in specified amounts makes a good
(a) Lewis acid (b) Bronsted acid
(c) strong acid (d) buffer

271. Soil structure in which the structural units have horizontal axes much longer than the vertical axes is
(a) blocky (b) columnar (c) prismatic (d) platy

272. This fertilizer is an acid forming fertilizer in the soil
(a) urea (b) calcium ammonium nitrate
(c) muriate of potash (d) sulphate of potash

273. Denitrification of plant available nitrate is favoured in
(a) neutral soils (b) waterlogged soils
(c) well aerated soils (d) loose and friable soils

274. Volatilization loss of soil nitrogen as ammonia is favoured in
(a) neutral soils (b) acid soils
(c) highly acidic soils (d) alkaline soils

275. According to FCO Specifications in India, neutral ammonium citrate soluble phosphate content (as P_2O_5 % by weight) of diammonium phosphate (DAP) must be minimum
(a) 16 (b) 41 (c) 46 (d) 48

276. The radioactive isotope P^{32} undergoes nuclear decay leading to the formation of
(a) S^{31} (b) S^{32}
(c) Si^{31} (d) Si^{32}

277. The term 'hue' denotes this feature of a colour
(a) dominant spectrum (b) intensity or brightness
(c) purity (d) wavelength

278. In alkaline soils, phosphorus is absorbed by plants in the form of
(a) H_3PO_4 (b) $(H_2PO_4)^-$ (c) $(HPO_4)^{2-}$ (d) $(PO_4)^{3-}$

279. Slickensides are often noticed in soils that belong to the Order
(a) Alfisol (b) Vertisol
(c) Histosol (d) Mollisol

280. 5th December of the year is celebrated as
(a) World Environment Day (b) World Soil Day
(c) World Maritime Day (d) World Wildlife Day

281. In Munsell's Soil Color chart, a numerical value of 10 for 'hue' represents
(a) black (b) grey (c) yellow (d) red

282. A plant in much use for indication of permanent wilting point is
(a) sunflower (b) safflower
(c) turnip (d) rapeseed

283. This country does have rich deposits of rock phosphate
(a) Iran (b) France
(c) Morocco (d) Indonesia

284. Individual crystals of kaolinite clay mineral are joined by
(a) covalent bonding (b) electrovalent bonding
(c) hydrogen bonding (d) van der Waals bonding

285. Microorganisms involved in the conversion of nitrite to nitrate are
(a) *Pseudomonas* (b) *Xanthomonas*
(c) *Nitrosomonas* (d) *Nitrobacter*

286. Nitrification is a process of
(a) oxidation (b) reduction
(c) hydration (d) hydrolysis

287. Mohs scale is followed to determine this property of minerals
(a) cleavage (b) refractive index
(c) fracture (d) hardness

288. The mineral that records the highest numerical value in Mohs scale is
(a) pyrites (b) rock phosphate
(c) diamond (d) limestone

289. Soils or parent material deposits formed by glaciers as the agency of transportation are called
(a) outwash (b) moraines
(c) fluvial (d) glacial

290. Essentiality of molybdenum for plant growth was first established by
(a) Sommer and Lipman (b) Brownell and Wood
(c) E. Gris (d) Arnon and Stout

291. Mohs scale followed to determine the property of hardnes of minerals ranges from
(a) 1 to 5 (b) 1 to 10 (c) 1 to 20 (d) 1 to 100

292. Essentiality of nitrogen as a plant nutrient was first established by
(a) Justus von Leibig (b) Theodore de Saussure
(c) D. J. Nicolas (d) Jean Baptiste Boussingault

293. Water is used in part as a source of hydrogen for the manufacture of ammonia in the fertilizer plant located at
(a) Nangal (b) Trombay (c) Sindri (d) Namrup

294. The term 'value' denotes this feature of a colour
(a) dominant spectrum (b) intensity or brightness
(c) purity (d) transparency

295. Conversion of urea to ammonium carbonate in the soil is a process of
(a) oxidation (b) reduction
(c) hydration (d) hydrolysis

296. In Soil Taxonomy, relatively young soils are classified under the Order
(a) Ultisol (b) Inceptisol
(c) Entisol (d) both Inceptisol and Entisol

297. An expression of Q (the volume flow of water in cm^3/second) in Poiseuille's law for flow of liquid through a narrow tube is
(a) $\frac{\pi R^4 \Delta P}{8\ \mu L}$ (b) $\frac{\pi R^2 \Delta P}{8\ \mu L}$
(c) $\frac{\pi R^4 \Delta P}{4\ \mu L}$ (d) $\frac{\pi R^4 \Delta P}{\mu L}$

298. Laterization soil formation process is favoured under
(a) warm humid climate (b) high rainfall
(c) basic parent material (d) all the three

299. The transporting agency in the formation of deposits of parent materials or soils referred as till and moraine is
(a) gravity (b) water (c) wind (d) ice

300. The terminal acceptor of electrons in the process of respiration by denitrifying bacteria is
(a) molecular oxygen (b) nitrate
(c) nitrite (d) elemental nitrogen

301. The method of composting developed by Hutchinson and Richards is
(a) ADCO process (b) Activated Compost methos
(c) Indore method (d) Coimbatore method

302. This source of hydrogen for the manufacture of ammonia for use in fertilizer plants is not preferred because of high energy requirement
(a) coal (b) naphtha (c) water (d) natural gas

303. The combustible gas produced in biogas plants is
(a) methane (b) propane (c) butane (d) octane

304. Thiourea is a
(a) coated urea (b) slow release urea
(c) slow acting urea (d) nitrification inhibitor

305. $CaHPO_4$ is a constituent of this fertilizer
(a) nitro phosphate (b) triple super phosphate
(c) single super phosphate (d) urea ammonium phosphate

306. This metal is a heavy metal
(a) lead (b) calcium (c) magnesium (d) barium

307. This substance is used as a primary standard acid for the standardization of prepared alkali solutions
   (a) potassium hydrogen *ortho*phosphate
   (b) potassium hydrogen phthalate
   (c) oxalic acid
   (d) all the three

308. Microorganisms for which the optimum temperature for growth is less than 10^0 C are called
   (a) mesophiles (b) thermophiles
   (c) psychrophiles (d) hyperthermophiles

309. Essentiality of iron for plant growth was first established by
   (a) Theodore de Saussure (b) C. Sprengel
   (c) E. Gris (d) Arnon and Stout

310. Microorganisms which can derive their carbon requirement from atmospheric CO_2 are called
   (a) heterotrophs (b) phototrophs
   (c) autotrophs (d) chemotrophs

311. Saline alkali soils are characterized by these soil characteristics
   (a) pH $<$ 8.5, EC $>$ 4 dS/m at 25^0C, ESP $<$ 15
   (b) pH $>$ 8.5, EC $>$ 4 dS/m at 25^0C, ESP $>$ 15
   (c) pH $>$ 8.5, EC $<$ 4 dS/m at 25^0C, ESP $>$15
   (d) pH $>$ 8.5, EC $>$ 4 dS/m at 25^0C, ESP $<$ 15

312. With decrease in soil acidity, the cation exchange capacity of silicate clays
   (a) increases (b) decreases
   (c) may increase or decrease (d) remains unaffected

313. Gypsum is a popular choice amendment for the reclamation of
   (a) acid soils (b) acid sulphate soils
   (c) alkali soils (d) calcareous soils

314. Individual crystals of this clay mineral are joined by hydrogen bonding
   (a) kaolinite (b) montmorillonite
   (c) illite (d) vermiculite

315. Soil profiles of the soils of this Order are developed from parent materials of volcanic origin
   (a) Aridisol (b) Andisol (c) Spodosol (d) Oxisol

316. A permafrost at a very low soil depth is associated with the soils of this Order
(a) Aridisol (b) Andisol (c) Spodosol (d) Gelisol

317. The refractive index of sea and ocean water
(a) increases with salinity (b) decreases with salinity
(c) increases with temperature (d) decreases with temperature

318. In remote sensing, the basic requirement of any sensor system is
(a) radiometric resolution (b) spectral resolution
(c) spatial resolution (d) all the three

319. Soil Taxonomy now in vogue has been devised by
(a) Food and Agriculture Organization
(b) International Society of Soil Science
(c) United States Department of Agriculture
(d) Soil Science Society of America

320. On dissolution in water, this fertilizer produces a nonionic solution
(a) borax (b) muriate of potash
(c) urea (d) potassium schoenite

321. Leghaemoglobin protects the enzyme nitrogenase from being degraded by
(a) molecular oxygen (b) ammonia
(c) water (d) nitrate

322. *Frankia* associated with biological nitrogen fixation are
(a) protozoa (b) fungi
(c) actinomycetes (d) algae

323. An indicator plant for affirming molybdenum deficiency in soils is
(a) rye (b) lucerne (c) sugar beet (d) cauliflower

324. With increase in pH, the CEC of humus
(a) increases (b) decreases
(c) may increase or decrease (d) remains unaltered

325. The most abundant genus of bacteria in soil is
(a) *Bacillus* (b) *Xanthomonas*
(c) *Pseudomonas* (d) *Escherichia*

326. The soil mass of a cylindrical core sampler (radius 6 cm, height 14 cm) on oven drying weighed 2.2176 kg. The bulk density of the soil is
(a) 1.2 g/cm^3 (b) 1.3 g/cm^3 (c) 1.4 g/cm^3 (d) 1.5 g/cm^3

327. Improper development of inflorescence in wheat is caused due to the deficiency of this nutrient
(a) iron (b) zinc
(c) molybdenum (d) boron

328. Amoozemeter is used to measure this property of soils
(a) soil moisture tension
(b) saturated hydraulic conductivity
(c) diffusion of gases
(d) thermal conductivity

329. High concentration of this anion helps suppressing the incidence of *Fusarium* wilt of a number of crop plants
(a) sulfite (b) nitrate (c) phosphate (d) sulphate

330. Pycnometer is used to measure this property of soils
(a) soil moisture tension
(b) saturated hydraulic conductivity
(c) density
(d) thermal conductivity

331. Flow of a fluid through a porous medium is quantified by
(a) Fick's law (b) Stokes' law
(c) Fourier's law (d) Darcy's law

332. Alkali soils are characterized by the following soil characteristics
(a) pH < 8.5, EC > 4 dS/m at 25^0C, ESP < 15
(b) pH > 8.5, EC > 4 dS/m at 25^0C, ESP > 15
(c) pH > 8.5, EC < 4 dS/m at 25^0C, ESP >15
(d) pH > 8.5, EC > 4 dS/m at 25^0C, ESP < 15

333. Soil amendment not to be used for the reclamation of acid soils is
(a) limestone (b) gypsum
(c) dolomite (d) basic slag

334. In nitrogen fixation carried out by soil microbes and by lightnings in the atmosphere, the kinds of chemical reaction suffered by the elemental dinitrogen of the soil and atmospheric air are
(a) reduction and oxidation (b) oxidation and reduction
(c) reduction and reduction (d) oxidation and oxidation

335. The CEC of montmorillonite, illite, vermiculite and kaolinite follows the following order
(a) montmorillonite > illite > vermiculite > kaolinite
(b) vermiculite > montmorillonite > illite > kaolinite
(c) vermiculite > montmorillonite > kaolinite > illite
(d) vermiculite > kaolinite > montmorillonite > illite

336. Magnesium, a secondary plant nutrient, is intimately associated with the functioning of
   (a) hemoglobin (b) cytochrome
   (c) ribosomes (d) ATP

337. This enzyme is an extra cellular enzyme
   (a) nitrogenase (b) urease
   (c) nitrate reductase (d) nitrite reductase

338. Soils or parent material deposits formed by gravity as the agency of transportation are called
   (a) alluvial (b) colluvial (c) fluvial (d) glacial

339. 'White bud' in maize is caused due to the deficiency of
   (a) copper (b) manganese
   (c) zinc (d) molybdenum

340. *Azolla pinnata* is a
   (a) halophyte (b) thallophyte
   (c) lycophyte (d) pteridophyte

341. DTPA, a widely used extractant for determining available micronutrient cations in soils, is a/an
   (a) tridentate ligand (b) octadentate ligand
   (c) hexadentate ligand (d) pentadentate ligand

342. Soils of this Order are intensely weathered and leached so as to reach a per cent base saturation less than 35
   (a) Aridisol (b) Spodosol (c) Ultisol (d) Andisol

343. pH meter measures a soil's
   (a) active acidity (b) reserve acidity
   (c) buffering capacity (d) potential buffering capacity

344. Granite is a rock of this class
   (a) acid igneous (b) basic igneous
   (c) stratified (d) metamorphic

345. Black soils are normally poor in
   (a) phosphorus (b) potassium
   (c) calcium (d) magnesium

346. A *kankar* layer is often noticed in
   (a) black soils (b) red soils
   (c) arid region soils (d) humid region soils

347. Black soils are normally rich in
(a) calcium (b) potassium
(c) magnesium (d) calcium, potassium and lime

348. These soils have high particle density
(a) black soils (b) red soils
(c) alluvial soils (d) laterite soils

349. Gneiss is a rock metamorphically formed from this rock
(a) granite (b) basalt (c) gabbro (d) andesite

350. The diameter of the micropores in soils is
(a) more than 0.06 mm (b) less than 0.06 mm
(c) 0.06 to 0.08 mm (d) 0.08 to 0.10 mm

351. The metamorphic rock produced from basalt, an igneous rock, is
(a) gneiss (b) schist (c) slate (d) quartzite

352. According to Soil Taxonomy, the Soil Orders to which the desert soils belong to are
(a) Alfisol and Aridisol (b) Entisol and Aridisol
(c) Entisol and Vertisol (d) Ultisol and Andisol

353. The soil often referred to as garden soil is
(a) alluvial soil (b) black soil
(c) red soil (d) muck soil

354. According to Soil Taxonomy, the soil Order Ultisol includes these soil types
(a) alluvial and black (b) black and red
(c) red and laterite (d) alluvial and red

355. Strength of soil crusts can be determined by
(a) tensiometer method
(b) modulus of rupture test
(c) pressure plate apparatus method
(d) pressure membrane apparatus method

356. The hardest of the minerals available in the earth's crust is
(a) quartz (b) coal (c) talc (d) diamond

357. Microorganisms involved in the conversion of ammonium to nitrite are the species of
(a) *Pseudomonas* (b) *Xanthomonas*
(c) *Nitrosomonas* (d) *Nitrobacter*

358. In Land Capability Classification, the Classes that are unsuitable for cultivation are
(a) Class I to Class IV (b) Class V to Class VIII
(c) Class I to Class V (d) Class VII and Class VIII

359. According to Soil Taxonomy, the soil Order Alfisol includes
(a) desert soils (b) black soils
(c) red soils (d) peat soils

360. This mineral contains both Al and Si
(a) sphene (b) goethite (c) powellite (d) topaz

361. The soil type that occupies the largest cultivable area in India is
(a) alluvial soil (b) black soil
(c) red soil (d) swampy soil

362. In local parlance, the term *Khadar* refers to
(a) fertile new alluvium (b) fertile old alluvium
(c) infertile new alluvium (d) infertile old alluvium

363. A perfect black body
(a) absorbs all radiations
(b) reflects all radiations
(c) absorbs only visible radiations
(d) reflects only visible radiations

364. ICAR-Central Soil Salinity Research Institute is located in
(a) New Delhi (b) Karnal
(c) Bhopal (d) Hyderabad

365. The order of binding of the anions to a positively charged surface follows the following order
(a) $PO_4^{3-} > SO_4^{2-} > Cl^- > NO_3^-$ (b) $SO_4^{2-} > PO_4^{3-} > Cl^- > NO_3^-$
(c) $Cl^- > PO_4^{3-} > NO_3^- > SO_4^{2-}$ (d) $NO_3^- > Cl^- > PO_4^{3-} > SO_4^{2-}$

366. The anion exchange capacity of the soil clays follows the following order
(a) 2:1 type clays > oxide and hydroxide clays > 1:1 type clays
(b) 1:1 type clays > 2:1 type clays > oxide and hydroxide clays
(c) 1:1 type clays > oxide and hydroxide clays > 2:1 type clays
(d) oxide and hydroxide clays > 1:1 type clays > 2:1 type clays

367. The capacity of the common cations to flocculate a colloidal system follows the following order
(a) $Al^{3+} > K^+ > Na^+ > Ca^{2+}$ (b) $Ca^{2+} > Al^{3+} > K^+ > Na^+$
(c) $Al^3 > Ca^{2+} > Na^+ > K^+$ (d) $Al^{3+} > Ca^{2+} > K^+ > Na^+$

368. The strength of binding of a cation to a negatively charged surface is a function of the charge of the cation and its
(a) atomic weight (b) atomic number
(c) solubility (d) radius of the hydrated cation

369. Ion exchange capacity soils is expressed in terms of
(a) m.e./100 g soil (b) cmol (p^+/e^-)/kg soil
(c) both the two (d) neither of the two

370. According to Soil Taxonomy, laterite soils do not belong to this soil Order
(a) Alfisol (b) Ultisol (c) Inceptisol (d) Oxisol

371. The diameter of the macropores in soils is
(a) more than 0.06 mm (b) less than 0.06 mm
(c) 0.04 to 0.06 mm (d) less than 0.04 mm

372. Organic carbon content of a soil multiplied by this factor gives an approximate estimate of organic matter content of the soil
(a) 0.58 (b) 1.21 (c) 1.724 (d) 2.29

373. The Standard Redox Potential of a redox system is denoted by the symbol
(a) E (b) P (c) E^0 (d) F

374. This layer lattice clay mineral is a limited expanding type clay mineral
(a) montmorillonite (b) kaolinite
(c) smectite (d) vermiculite

375. Gabbro is a rock of this class
(a) acid igneous (b) basic igneous
(c) stratified (d) metamorphic

376. Honeycomb soil structure is noticed in
(a) alluvial soils (b) black soils
(c) alkali soils (d) laterite soils

377. Soil crusting poses a problem in
(a) sandy soils (b) sandy clay soils
(c) clay soils (d) silty clay soils

378. In India, currently nano urea is being manufactured and marketed by
(a) Fertilizer Corporation of India
(b) ZuariAgro Chemicals
(c) Fertilizers and Chemicals Travancore
(d) Indian Farmers Fertilizer Cooperative Ltd (IFFCO)

379. On addition of organic matter to a soil, the bulk density of the soil
   (a) increases (b) decreases
   (c) may increase or decrease (d) remains unchanged
380. The clay mineral that causes cracks in black soils upon drying is
   (a) kaolinite (b) illite
   (c) montmorillonite (d) chlorite
381. Surface crusts made by algae, fungi, lichens, bacteria and mosses are crusts of the following kind
   (a) structural (b) erosion
   (c) depositional (d) cryptogamic
382. ICAR-Indian Institute of Soil and Water Conservation is located in
   (a) Karnal (b) Dehradun
   (c) Bhopal (d) Hyderabad
383. The headquarter of International Union of Soil Sciences is located in
   (a) Washington (b) Madison
   (c) Vienna (d) New Delhi
384. Ribbon behaviour test with a moist bolus of soil is performed to have an approximate estimate of
   (a) soil texture (b) soil structure
   (c) soil strength (d) soil thermal behaviour
385. *Karewas*, a unique soil/soil deposit found in Jammu and Kashmir is a deposit of this type
   (a) dune (b) lacustrine (c) moraine (d) till
386. The most widespread of the different soil types in Deccan plateau is
   (a) black soils (b) red soils
   (c) laterite soils (c) alluvial soils
387. Nickel is a metallic cofactor of the enzyme
   (a) nitrogenase (b) phosphatase
   (c) sulfatase (d) urease
388. Transfer of heat from one place to another without the help of an intervening or supporting medium is called
   (a) dissipation (b) conduction
   (c) convection (d) radiation
389. The soils of the Gangetic plains of India are mostly
   (a) black soils (b) red soils
   (c) laterite soils (d) alluvial soils

390. A practical way of reducing NH_4^+ fixation in the inter lattice space of 2:1 type clay minerals following the application of ammonical fertilizers is prior application of potassic fertilizer because
(a) NH_4^+ is more soluble than K^+
(b) NH_4^+ is held more tightly than K^+ on exchange complexes
(c) ionic radii of hydrated NH_4^+ and K^+ are matching
(d) atomic masses of NH_4^+ and K^+ are matching

391. With increase in the content of 2:1 type clay minerals, the degree of NH_4^+ fixation by soil
(a) increases (b) decreases
(c) may increase or decrease (d) no relationship between the two

392. Marble is a rock of this class
(a) igneous (b) sedimentary
(c) metamorphic (d) volcanic

393. For the determination of available potassium content of soils, the extractant used is
(a) neutral 1 N ammonium acetate
(b) Olsen's reagent
(c) Bray's No. 1 reagent
(d) Morgan's extractant

394. Keen and Rackzowski's boxes are used for the determination of a soil's
(a) structure
(b) texture
(c) bulk density and particle density
(d) thermal conductivity

395. Bouyoucos, a soil physicist, developed a special type of this instrument for carrying out the particle size analysis of soils
(a) pycnometer (b) hydrometer
(c) thermometer (d) viscometer

396. The capacity of the cations Ca^{2+}, Mg^{2+}, K^+ and Na^+ to keep a colloidal system in a state of dispersion (deflocculated state) follows the following order
(a) $Mg^{2+} > K^+ > Na^+ > Ca^{2+}$ (b) $Ca^{2+} > Mg^{2+} > K^+ > Na^+$
(c) $Ca^{2+} > Mg^{2+} > Na^+ > K^+$ (d) $Na^+ > K^+ > Mg^{2+} > Ca^{2+}$

397. The cation exchange capacity of the soil clays follows the following order
   (a) 2:1 type clays > oxide and hydroxide clays > 1:1 type clays
   (b) 1:1 type clays > 2:1 type clays > oxide and hydroxide clays
   (c) 2:1 type clays > 1:1 type clays > oxide and hydroxide clays
   (d) 1:1 type clays > oxide and hydroxide clays > 2:1 type clays

398. A simple method for the determination of organic carbon content of soils was developed by
   (a) Kjeldahl (b) Walkley and Black
   (c) Subbaiah and Asija (d) Lindsay and Norvel

399. Nessler's reagent is used in the detection and determination of
   (a) ammonium (b) nitrate
   (c) nitrate (d) amines

400. The diameter of silt particles according to ISSS classification system is
   (a) 0.2 to 2 mm (b) 0.02 to 0.2 mm
   (c) 0.002 to 0.02 mm (d) less than 0.002 mm

401. The most abundant of all the elements in the earth's crust is
   (a) oxygen (b) silicon
   (c) iron (d) aluminium

402. Clay minerals are
   (a) primary minerals (b) secondary minerals
   (c) accessory minerals (d) essential minerals

403. With regard to genesis, limestone is a sedimentary rock of this type
   (a) residual (b) mechanical
   (c) chemical (d) organic

404. The mineral that accounts for 50 to 90% of the mass of the sand and coarse silt fractions is
   (a) quartz (b) feldspars (c) micas (d) olivines

405. If the top of the structural units in a columnar structure are relatively angular and flat horizontally, the soil structure is designated as
   (a) blocky (b) columnar
   (c) prismatic (d) platy

406. This metal is a key component of the chemical structure of hemoglobin
   (a) iron (b) cobalt (c) manganese (d) magnesium

407. Formation of urea from ammonium carbamate in the manufacture of urea is a process of

(a) oxidation (b) hydrolysis
(c) hydration (d) dehydration

408. This metal is a key component of the Photo System II of plants

(a) iron (b) copper (c) manganese (d) zinc

409. The micronutrient that now finds almost a regular inclusion in the fertilizer schedule in Indian soils is

(a) iron (b) copper (c) manganese (d) zinc

410. The unit for expressing the viscosity of fluids is

(a) dBel (b) Siemens (c) Poise (d) Mho

411. 'Mottle leaf' in citrus is caused due to the deficiency of

(a) iron (b) copper (c) manganese (d) zinc

412. Soils or parent material deposits formed by wind as the agency of transportation are called

(a) outwash (b) aeolian (c) fluvial (d) glacial

413. DRIS as a diagnostic approach in soil fertility evaluation was developed by

(a) Arnon (b) Morgan
(c) Lindsay and Norvel (d) Beaufils

414. The concept of Critical Limits of plant nutrients for assessing their deficiencies in soils was developed by

(a) Arnon (b) Morgan
(c) Cate and Nelson (d) Beaufils

415. Boron is absorbed by plants in the form of

(a) H_3BO_3 (b) $(H_2BO_3)^-$ (c) $(HBO_3)^{2-}$ (d) All the three

416. In waterlogged soils, sulphur exists predominantly as

(a) $(SO_4)^{2-}$ (b) $(SO_3)^{2-}$ (c) S^{2-} (d) $(S_2O_3)^{2-}$

417. Acid equivalent of the following fertilizers follows the order:

(a) ammonium sulphate > calcium ammonium nitrate > urea > anhydrous ammonia
(b) Urea > ammonium sulphate > calcium ammonium nitrate > anhydrous ammonia
(c) anhydrous ammonia > ammonium sulphate > urea > calcium ammonium nitrate
(d) ammonium sulphate > anhydrous ammonia > urea > calcium ammonium nitrate

418. Pyrites are iron
(a) sulphides (b) sulphites
(c) sulphates (d) thiosulphates

419. In low pH acid soils, soluble phosphorus is fixed as insoluble
(a) phosphoric acid (b) zinc phosphate
(c) monocalcium phosphate (d) aluminium phosphate

420. Pressure Plate Apparatus is used to measure this property of soils
(a) soil moisture retention
(b) saturated hydraulic conductivity
(c) density
(d) thermal conductivity

421. Bonemeal is a rich source of
(a) calcium and potassium (b) calcium and phosphorus
(c) calcium and iron (d) calcium and magnesium

422. Kjeldahl method is used for the determination of soil's
(a) total nitrogen (b) available nitrogen
(c) ammonium nitrogen (d) nitrate nitrogen

423. The instrument in use for the determination of micronutrient cations in soils is
(a) colorimer
(b) flame photometer
(c) spectrometer
(d) atomic absorption spectrophotometer

424. The equivalent weight of $MnSO_4$ (molecular weight 151.0) in its oxidation to $KMnO_4$ is
(a) 151.0 (b) 75.5 (c) 30.2 (d) 25.2

425. Denitrification is process of
(a) oxidation (b) reduction
(c) hydration (d) hydrolysis

426. The ionic form of nitrogen toxic to plants is
(a) NO_2^- (b) NO_3^- (c) NH_4^+ (d) $-NH_3^+$

427. The reagent that acts as a catalyst in the digestion of soil and plant samples by Kjeldahl's method of total nitrogen determination is
(a) potassium sulphate (b) cupric sulphate
(c) salicylic acid (d) Devarda's alloy

428. The mixed indicator used in the back titration of ammonium borate to boric acid against a standard acid is made of bromocresol green and
(a) phenolphthalein (b) bromocresol orange
(c) litmus (d) methyl red

429. Soil amendment that is not to be used for the reclamation of alkali soils is
(a) limestone (b) gypsum
(c) elemental sulphur (d) iron pyrites

430. Nitrification inhibitors inhibit the following step of the process of nitrification
(a) $NH_3 \rightarrow NH_4^+$ (b) $NH_4^+ \rightarrow NO_2^-$
(c) $NH_4^+ \rightarrow NO_3^-$ (d) $NO_2^- \rightarrow NO_3^-$

431. Rich deposits of guano are found in these countries
(a) Indonesia and Malayasia (b) Morocco and Jordon
(c) Peru and Chile (d) France and England

432. Coal is a rock of this class
(a) igneous (b) sedimentary
(c) metamorphic (d) volcanic

433. With increase in the compaction of a soil, the particle density of the soil
(a) decreases (b) increases
(c) may increase or decrease (d) remains unchanged

434. In soil analysis laboratories, molybdenum blue method is followed for the determination of soil
(a) nitrogen (b) phosphorus (c) sulphur (d) boron

435. Spectronic 20, a much used analytical instrument in soil analysis laboratories, is a variant of
(a) absorption spectrophotometer
(b) flame photometer
(c) EC Bridge
(d) gas chromatograph

436. A pH meter does have two electrodes viz. a reference electrode and a/an
(a) standard hydrogen electrode
(b) calomel electrode
(c) silver-silver chloride electrode
(d) indicator electrode

437. Aluminium chloride is an acid according to the acid base concept of
   (a) Arrhenius (b) Lowry
   (c) Bronsted (d) Lewis
438. Soils of high rainfall areas are normally
   (a) acidic (b) neutral (c) alkaline (d) calcareous
439. The fertilizer with the highest numerical value of acid equivalent is
   (a) ammonium sulphate (b) ammonium nitrate
   (c) nitrophosphate (d) anhydrous ammonia
440. Ratio law was formulated by
   (a) Sorenson (b) Jenny
   (c) Schofield (d) Lipman
441. A property not characteristic of colloidal particles is their
   (a) high surface area (b) adsorption behavoiur
   (c) possessing electrical charge (d) coarse particle size
442. Boron deficiency is relatively uncommon in
   (a) sandy soils
   (b) clayey soils
   (c) alkaline soils
   (d) soils formed from granitic parent material
443. *Rhizobium* species specific to soybean as the host plant is
   (a) *R. japonicum* (b) *R. lupini*
   (c) *R. leguminosarum* (d) *R. meliloti*
444. Collision of molecules and passing on their kinetic energy to adjacent molecules is associated with the transfer of heat through the process of
   (a) dissipation (b) conduction
   (c) convection (d) radiation
445. The mineral that records the least numerical value of hardness in Mohs scale is
   (a) pyroxene (b) biotite
   (c) mica (d) talc
446. Theory of spectrophotometry is based upon the principles of
   (a) Snell's law (b) Kirchoff's law
   (c) Lambert-Beer's law (d) Beckman's law
447. The unit for expressing the Residual Sodium Carbonate (RSC) value of irrigation water is
   (a) mg/100 ml (b) mg/100 g
   (c) m.e./100 ml (d) m.e./L

448. Soils or parent material deposits referred as aeolian are formed by the following agency of transportation

(a) water (b) wind (c) ice (d) gravity

449. Irrigation water with this numerical value of Residual Sodium Carbonate (RSC) is generally safe for irrigation

(a) less than 1.25 (b) 1.25 to 2.50
(c) 2.50 to 3.50 (d) more than 4.50

450. A guano forming sea bird is

(a) pelican (b) gannets
(c) albatross (d) all the three

451. Availability of phosphorus is maximum at this soil pH

(a) 5 to 6 (b) 6 to 7 (c) 7 to 8 (d) more than 8

452. Heat flow in soils is described by

(a) Fick's law (b) Stokes' law
(c) Fourier's law (d) Darcy's law

453. Decomposability of the constituents of soil organic matter follows the sequence is

(a) proteins > cellulose > sugars > lignins
(b) cellulose > proteins > lignins > sugars
(c) lignins > sugars > cellulose > proteins
(d) sugars > proteins > cellulose > lignins

454. The C:N ratio of saw dust is

(a) 10:1 (b) 50:1 (c) 100:1 (d) 400:1

455. Nano urea currently being marketed in India is a formulation of this kind

(a) liquid (b) prills (c) granules (d) super granules

456. Thomas Way, an early Soil Scientist, first reported the

(a) biological nitrogen fixation by legumes
(b) colloidal behaviour of silicate clays
(c) cation exchange property of soils
(d) Law of Minimum in plant nutrition

457. An important geological feature of Deccan plateau is

(a) vast tracts of black soils produced from basaltic rocks
(b) vast tracts of alluvial soils transported by water
(c) vast tracts of sandy soils produced from igneous rocks
(d) vast tract of lateritic soils produced from laterite rocks

458. Most abundant of the different classes of rocks is
    (a) igneous (b) sedimentary
    (c) metamorphic (d) clastic

459. A higher value of Coefficient of Linear Extensibility (COLE) of a soil is indicative of its higher content of
    (a) expanding type clay minerals
    (b) nonexpanding type clay minerals
    (c) organic matter
    (d) fine sand and silt

460. Robinson's pipette method is used for the determination of soil
    (a) particle density (b) bulk density
    (c) water holding capacity (d) particle size analysis

461. Salicylic acid is used in the determination of total nitrogen by Kjeldahl method for the inclusion of
    (a) ammonium nitrogen (b) nitrite nitrogen
    (c) nitrate nitrogen (d) amino nitrogen

462. Nitrification inhibitors are inhibitory to the growth of the bacteria
    (a) *Nitrosomonas* (b) *Nitrobacter*
    (c) *Pseudomonas* (d) *Xanthomonas*

463. The SAR of an irrigation water source is 12.5. The irrigation water is rated as
    (a) S1 Class. Low sodium hazard
    (b) S2 Class. Medium sodium hazard
    (c) S3 Class. High sodium hazard
    (d) S4 Class. Very high sodium hazard

464. This potassic fertilizer is manufactured in India
    (a) muriate of potash (b) sulphate of potash
    (c) nitre (d) potassium schoenite

465. The fertilizer SUPHALA is
    (a) monoammonium phosphate (b) diammonium phosphate
    (c) urea ammonium phosphate (d) N-P-K complex

466. The terms epipedon and endopedon in Soil Taxonomy differ with regard to their
    (a) depth in the soil (b) physical properties
    (c) minerology (d) microbial population

467. Ammonium sulphate is produced as a byproduct in
(a) iron and steel industry (b) cement industry
(c) paper industry (d) thermal power plants

468. Hornmeal is a rich source of
(a) nitrogen (b) phosphorus (c) calcium (d) sulphur

469. Pit, Covered Pit and Heap method are different methods for
(a) collection of FYM (b) storage of FYM
(c) preparation of compost (d) preparation of poudrette

470. Of all the organic manures, the one that has the highest availability of nitrogen is
(a) fish meal (b) horn meal (c) bone meal (d) blood meal

471. Slate, a metamorphic rock, is produced from this rock
(a) soapstone (b) sandstone
(c) dolomite (d) shale

472. The green manuring crop that is also a minor pulse crop and does have considerable fodder value is
(a) *Sesbania aculeata* (b) *Crotolaria juncea*
(c) *Phaseolus trilobus* (d) *Tephrosia purpurea*

473. The mixed indicator used in the back titration of ammonium borate to boric acid against a standard acid is a mix of methyl red and
(a) phenolphthalein (b) bromocresol orange
(c) litmus (d) bromocresol green

474. Biotite is a mineral under the following class
(a) feldspars (b) olivines
(c) amphiboles (d) micas

475. Siderite is a mineral of the following class
(a) oxide (b) sulphate (c) sulphite (d) carbonate

476. This mineral contains water of crystallization as an integral part of the molecule
(a) calcite (b) gypsum (c) limestone (d) dolomite

477. The equivalent weight of gypsum is
(a) 68.1 (b) 136.2 (c) 86.1 (d) 172.2

478. Charcoal like substance made by burning organic material from agricultural and forestry wastes in a controlled pyrolysis process is
(a) Biomass (b) Biosuper (c) Biochar (d) Biota

479. Actinomycetes for their growth prefer a medium that is
(a) neutral (b) slightly acidic
(c) slightly alkaline (d) strongly alkaline

480. Rate of diffusion of gases between two systems is described by
(a) Fick's law (b) Stokes' law
(c) Fourier's law (d) Poiseuille's law

481. Fungi for their growth prefer a medium that has a pH of
(a) 4.5 to 6.5 (b) 6.5 to 7.5
(c) 7.5 to 8.5 (d) more than 8.5

482. One of the following rocks is not a polymineralitic rock
(a) quartzite (b) basalt (c) granite (d) diorite

483. Soil crusts may be
(a) physical crusts (b) chemical crusts
(c) biological crusts (d) all the three

484. Brownian movement, a characteristic of colloidal solutions, is a property of this kind
(a) electrical (b) mechanical
(c) optical (d) magnetic

485. The reagent added for elevating the boiling point of concentrated sulphuric acid in the digestion of samples by Kjeldahl's method of total nitrogen determination is
(a) potassium sulphate (b) cupric sulphate
(c) salicylic acid (d) Devarda's alloy

486. Rice because of its prolonged requirement of submerged conditions grows well in
(a) light textured soils (b) heavy textured soils
(c) well aerated soils (d) well drained soils

487. Clumps and fragments of rocks are more abundant in this soil horizon
(a) A horizon (b) B horizon (c) C horizon (d) D horizon

488. Muscovite is a mineral of the following class
(a) feldspars (b) olivines
(c) amphiboles (d) micas

489. The seeds of this green manuring crop require scarification for facilitating easy germination
(a) dhaincha (b) indigo
(c) wild indigo (d) sunnhemp

490. Hausmannite is a mineral containing
(a) titanium (b) manganese (c) copper (d) iron

491. Rutile is a mineral of the following type
(a) oxide (b) sulphate (c) sulphite (d) carbonate

492. This soil horizon contains more humus than any of the other horizons
(a) A horizon (b) B horizon
(c) C horizon (d) D horizon

493. A soil becomes more adhesive, cohesive and plastic with its higher content of
(a) fine sand (b) coarse sand
(c) silt (d) clay

494. Water retention capacity of soils is directly related to their content of
(a) fine sand (b) coarse sand
(c) silt (d) clay

495. Profiles of the soils of this Order are still under development
(a) Alfisol (b) Vertisol (c) Mollisol (d) Inceptisol

496. Micas are layer lattice minerals of the following type
(a) 1:1 (b) 2:1 (c) 2:1:1 (d) all the three

497. Byre method is a method concerned with the
(a) collection of FYM (b) storage of FYM
(c) preparationofcompost (d) preparationof poudrette

498. Tyndall effect, a property characteristic of colloidal solutions, is a property of this kind
(a) electrical (b) mechanical
(c) optical (d) sedimentation

499. India's first fertilizer manufacturing plant was established at
(a) Alwaye in Kerala (b) Ranipet in Tamil Nadu
(c) Sindri in Jharkhand (d) Kalol in Gujarat

500. The fertilizer the consumption of which is the highest in India
(a) ammonium sulphate (b) urea
(c) calcium ammonium nitrate (d) diammonium phosphate

501. One among the following fertilizers is commercially available in granular form is
(a) ammonium nitrate (b) single super phosphate
(c) ammonium chloride (d) diammonium phosphate

502. The second most widely used fertilizer in India is
(a) ammonium sulphate (b) urea
(c) calcium ammonium nitrate (d) diammonium phosphate

503. Ammonium polyphosphate is a
(a) crystalline fertilizer (b) coated fertilizer
(c) granulated fertilizer (d) liquid fertilizer

504. If the mean annual soil temperature is $< 0^0$ C, usually with permafrost, the Soil Temperature Regime (STR) for use in Soil Taxonomy is referred as
(a) Frigid (b) Gelic (c) Pergelic (d) Cryic

505. PEC process is used in the manufacture of
(a) NPK complexes (b) nitrophosphates
(c) urea ammonium phosphate (d) ammonium phosphate sulphate

506. A radioisotope of much concern for increasing levels of its presence in soils is
(a) ^{3}H (b) ^{32}S (c) ^{39}K (d) ^{137}Cs

507. This mineral does not contain iron
(a) pyrites (b) siderite (c) ilmenite (d) sphalerite

508. Black colour of black soils is due to the presence of
(a) hematite (b) limonite
(c) muscovite (d) titaniferous magnetite

509. Chlorofluorocarbons are potent environment pollutant gases because they
(a) increase pollution load of the soil
(b) contaminate the lithosphere of the earth
(c) pollute the underground hydrosphere of the earth
(d) deplete the protective ozone layer cover of the earth

510. The state of India which has the maximum fertilizer manufacturing facility in terms of tonnage is
(a) Uttar Pradesh (b) Tamil Nadu
(c) Maharashtra (d) Gujarat

511. This fertilizer is a straight fertilizer
(a) monoammonium phosphate (b) diammonium phosphate
(c) nitrophosphate (d) single super phosphate

512. Nitrophosphates are manufactured by
(a) floatation process (b) Manheim furnace process
(c) Den process (d) ODDA process

513. In the manufacture of urea, the latter is produced by the dehydration of
(a) carbamide (b) ammonium carbamate
(c) ammonium carbonate (d) ammonium bicarbonate

514. *Nostoc* genera are
(a) bacteria (b) fungi
(c) actinomycetes (d) algae

515. Solubilization of insoluble rock phosphate involves
(a) hydrolysis (b) oxidation
(c) reduction (d) carbonation

516. The soils with this endopedon are intensely weathered
(a) calcic (b) natric (c) histic (d) oxic

517. One among the following micronutrient elements is not polyvalent
(a) iron (b) manganese
(c) zinc (d) molybdenum

518. The scientist who contributed much to the modern USDA Soil Taxonomy is
(a) C F Marbut (b) V V Dokuchaev
(c) Guy D Smith (d) Hans Jenny

519. It is not advisable to apply potassic fertilizers to coarse textured soils just before the onset of rains because of their loss by
(a) fixation (b) leaching
(c) mineralization (d) immobilization

520. In Soil Taxonomy, an identifying geographic name such as that of a river, town, area etc. is used in naming a
(a) Subgroup (b) Family
(c) Series (d) Phase

521. If the present radioactivity of a plant sample containing the radioisotope ^{32}P ($t_{1/2}$ is 14.2 days) is 40000 dps, the radioactivity of the sample after 70 days shall be close to
(a) 20000 dps (b) 5000 dps
(c) 2500 dps (d) 1250 dps

522. The most important hydrocarbon constituent of natural gas is
(a) methane (b) ethane
(c) propane (d) butane

523. If the mean annual soil temperature (MAST) is > 15^0 C but < 22^0 C, the Soil Temperature Regime (STR) for reference in Soil Taxonomy is referred as

(a) Frigid (b) Thermic (c) Mesic (d) Cryic

524. Microorganisms which can derive their energy requirement from the oxidation of organic compounds are called

(a) heterotrophs (b) phototrophs
(c) autotrophs (d) chemoautotrophs

525. The relationship of the decrease in the intensity of the transmitted radiation with an increase in the concentration of radiation absorbing substance in a solution is quantified by

(a) Fourier's law (b) Beer's law
(c) Bouguer's law (d) Darcy's law

526. The usually preferred concentration of urea for its foliar application is

(a) 0.05 to 0.10% (b) 0.10 to 0.50%
(c) 0.50 to 2.00% (d) 2.00 to 5.00%

527. The fertilizer manufactured from caprolactam manufacturing plants is

(a) ammonium sulphate (b) urea
(c) nitrophosphate (d) diammonium phosphate

528. A simple method for the laboratory determination of Lime Requirement of acid soils was described by

(a) Schoonover (b) Mehlich
(c) Morgan (d) Shoemaker

529. Nitrogenase, the nitrogen fixing enzyme, is protected from being degraded by molecular oxygen by

(a) porphyrin (b) siroheme
(c) ferredoxin (d) leghemglobin

530. The fertilizer for which FCO Specifications for Fertilizers in India permit moisture content upto a maximum of 12 per cent by weight is

(a) single super phosphate (b) diammonium phosphate
(c) nirophosphates (d) ammonium polyphosphate

531. Super phosphoric acid is used in the manufacture of

(a) triple super phosphate (b) diammonium phosphate
(c) urea ammonium phosphate (d) ammonium polyphosphate

532. If the mean annual soil temperature (MAST) is > 0^0 C, but < 8^0 C, the difference between mean summer and winter soil temperature is > 5^0 C, and there is cool summer temperatures, the Soil Temperature Regime (STR) for reference in Soil Taxonomy is referred as
(a) Frigid (b) Gelic (c) Pergelic (d) Cryic

533. Sphene is a mineral containing
(a) titanium (b) iron (c) copper (d) manganese

534. This fertilizer is a complex fertilizer
(a) monoammonium phosphate (b) urea
(c) calcium ammonium nitrate (d) single superphosphate

535. Nitrophosphates are manufactured by
(a) floatation process (b) Manheim furnace process
(c) Den process (d) PEC process

536. If the mean annual soil temperature (MAST) is > 0^0 C but < 8^0 C, the difference between mean summer and mean winter soil temperature is > 5^0 C and there is warm summer temperatures, the Soil Temperature Regime (SMR) for reference in Soil Taxonomy is referred as
(a) Frigid (b) Gelic (c) Pergelic (d) Cryic

537. This element is intimately associated with functioning of the enzyme nitrate reductase
(a) boron (b) copper
(c) manganese (d) molybdenum

538. The purpose of adding potassium sulphate in the digestion of soil and plant samples by Kjeldahl's method of total nitrogen determination is
(a) catalyzing the digestion process
(b) elevation of the boiling point of concentrated sulphuric acid
(c) inclusion of nitrate nitrogen
(d) to prevent bumping and frothing of the digestion mixture

539. A radio isotope of carbon (atomic number 6) of much use in soil research is
(a) C^{11} (b) C^{12} (c) C^{13} (d) C^{14}

540. In Soil Taxonomy, highly weathered soils are classified under the Order
(a) Inceptisol (b) Andisol
(c) Oxisol (d) Aridisol

541. Shoemaker described a simple method for the laboratory determination of this characteristic of soils
   (a) Cation Exchange Capacity
   (b) Exchangeable Sodium Per centage
   (c) Lime Requirement of acid soils
   (d) Gypsum Requirement of alkali soils

542. *Anabaena* associated with the process of biological nitrogen fixation are
   (a) bacteria (b) fungi
   (c) actinomycetes (d) algae

543. A solution of 10 parts per million (ppm) of iron contains
   (a) 10 mg iron in 100 ml solution
   (b) 10 mg iron in 1 L solution
   (c) 10 dg ironin100 ml solution
   (d) 10 dg iron in 1 L solution

544. These soils are referred to as white alkali soils because of a white encrustation of salts on the soil surface
   (a) acid soils (b) alkali soils
   (c) saline alkali soils (d) saline soils

545. Petrology is the study of formation, composition, structure, properties, and such aspects of
   (a) petroleum products (b) rocks
   (c) minerals (d) underground hydrosphere

546. The formation of a black colour on the soil surface is often noticed in
   (a) acid soils (b) alkali soils
   (c) saline soils (d) calcareous soils

547. A soil profile is a physical exposure of a soil along
   (a) horizontally (b) vertically
   (c) laterally (d) longitudinally

548. Paddy husk is composed of principally
   (a) hemicellulose and crude proteins
   (b) waxes and resins
   (c) silica and lignins
   (d) alumina and fibre proteins

549. The wavelength of the following electromagnetic radiations decreases in the order
   (a) X-ray < infrared < visible < ultraviolet
   (b) visible < infrared < ultraviolet < X-ray
   (c) γ-ray < ultraviolet < visible < radio waves
   (d) infrared < visible < ultraviolet < X-ray

550. The profile of a soil is a historical record of the
   (a) soil forming processes (b) parent material
   (c) soil biological activity (d) soil physical properties

551. The first fertilizer manufacturing plant of post independent India was established at
   (a) Alwaye in Kerala (b) Ranipet in Tamil Nadu
   (c) Sindri in Jharkhand (d) Phulpur in Uttar Pradesh

552. Lime Requirement of an acid soil is related to the soil's
   (a) texture (b) cation exchange capacity
   (c) clay content (d) all the three

553. The second most abundant element in the earth's crust, next only to oxygen, is
   (a) hydrogen (b) silicon
   (c) iron (d) aluminium

554. Diagnostic subsurface horizon (endopedon) characterized by accumulation and cementation of insoluble limestone and dolomite is
   (a) Gypsic (b) Salic (c) Calcic (d) Petrocalcic

555. Marble, a metamorphic rock, is produced from the rock
   (a) soapstone (b) sandstone
   (c) limestone (d) shalestone

556. An example of a carbonate mineral is
   (a) magnesite (b) siderite
   (c) smithsonite (d) all the three

557. The theory proposed by lundegardh for the absorption of ionic nutrients by plants is known as
   (a) Contact Exchange theory
   (b) Facilitated Diffusion of Nutrients
   (c) Passive Uptake of Nutrients against a concentration gradient
   (d) Cytochrome Pump theory

558. Cobalt, a beneficial plant nutrient for higher plants, is essential for
   (a) nitrogen fixing microbes
   (b) Azolla
   (c) phosphate solubilizing bacteria
   (d) Mycorrhizae

559. The practice that may aggravate soil crusting is
   (a) liberal addition of organic matter
   (b) application of single super phosphate
   (c) surface mulching
   (d) shallow ploughing

560. A primary standard acid for the standardization of a prepared NaOH solution is
   (a) hydrochloric acid (b) nitric acid
   (c) sulphuric acid (d) benzoic acid

561. Rock phosphates of Indian origin are not suitable for the manufacture of phosphatic fertilizers because of their
   (a) low water solubility (b) low phosphorus content
   (c) presence of impurities (d) presence of silica

562. The purpose of adding gypsum in the collection and storage of farm yard manure is to
   (a) promote microbial activity
   (b) increase calcium content of the manure
   (c) decrease losses of nitrogen from FYM
   (d) improve physical state of the manure

563. A simple method for the laboratory determination of Gypsum Requirement of acid soils was described by
   (a) Schoonover (b) Mehlich
   (c) Morgan (d) Shoemaker

564. Rock phosphate can be applied as a phosphatic fertilizer along with
   (a) gypsum
   (b) elemental sulphur
   (c) phosphate solubilizing bacteria
   (d) all the three

565. *Eisenia foetida* is a
   (a) nitrogen fixing alga (b) compost earthworm
   (c) guano forming bird (d) sulphur oxidizing bacteria

566. With increase in pH, the cation exchange capacity of humus and clays
(a) increases (b) decreases
(c) remains unchanged (d) may increase or decrease

567. According to the norms of the Soil Health Card Scheme of the Government of India, the grid area for drawing soil samples in the irrigated areas is
(a) 1 ha (b) 2.5 ha (c) 10 ha (d) 25 ha

568. For reference in Soil Taxonomy, soils that remain saturated with water for most part of the year as in tidal marsh are assigned the following Soil Moisture Regime (SMR)
(a) Udic (b) Aquic (c) Peraquic (d) Torric

569. Lowry and Bronsted proposed a theory for the
(a) nature of acids and bases
(b) mechanism of action of buffers
(c) degree of dissociation of weak acids and bases
(d) solubility product principles

570. A primary standard base for the standardization of a prepared HCl solution is
(a) sodium hydroxide (b) potassium hydroxide
(c) sodium carbonate (d) ammonia

571. Pampas grasslands are widespread in
(a) South America (b) Africa
(c) Siberia (d) Australia

572. Meghdoot, a joint initiative of the Government of India, Indian Meteorological Department and Indian Council of Agricultural Research for helping farmers in weather based farming, is a
(a) Call Centre (b) Mobile Application (App)
(c) database (d) satellite

573. Gypsum Requirement of an alkali soil is related to the soil's
(a) texture (b) cation exchange capacity
(c) clay content (d) all the three

574. An epipedon common in soils is
(a) Ochric (b) Argillic (c) Natric (d) Oxic

575. If the Residual Sodium Carbonate (RSC) value of a water source is 0.80 m.e./L, the water is categorized as
   (a) generally safe for irrigation
   (b) marginally safe for irrigation
   (c) usually unsuitable for irrigation without amendment
   (d) unsuitable for irrigation

576. Effectiveness of rock phosphate for its direct application as a phosphatic fertilizer in acid soils increases with its
   (a) purity
   (b) fine particle size
   (c) intimate mixing with the soil
   (d) all the three

577. Ammonium sulphate is produced as a byproduct in the industry manufacturing
   (a) cement (b) petroleum
   (c) caprolactam (d) thermal power

578. These two elements alone account for nearly 75% of the mass of the earth's crust
   (a) oxygen and hydrogen (b) oxygen and silicon
   (c) silicon and iron (d) silicon and aluminium

579. The number of soil test parameters in the Soil Health Card Scheme of the Government of India is
   (a) 8 (b) 10 (c) 12 (b) 14

580. In Soil Taxonomy, Salic, Sombric and Glossic are different
   (a) peds (b) pedons (c) epipedons (d) endopedons

581. Microorganisms for which the optimum temperature for growth is between 20 to 40^0 C are called
   (a) mesophiles (b) thermophiles
   (c) psychrophiles (d) hyperthermophiles

582. Manufacture of phosphatic fertilizers from rock phosphate requires
   (a) strong heat treatment (b) strong acid treatment
   (c) either of the two (d) neither of the two

583. The purpose of adding cupric sulphate in the digestion of soil and plant samples by Kjeldahl's method of total nitrogen determination is
   (a) catalyzing the digestion process
   (b) elevation of the boiling point of concentrated sulphuric acid
   (c) inclusion of nitrate nitrogen
   (d) to prevent bumping and frothing of the digestion mixture

584. In Soil Taxonomy, a diagnostic soil horizon characterized by the cementation by silica to form a subsurface hardpan is referred to by the term
   (a) Duripan (b) Fragipan (c) Plaggen (d) Sombric

585. The particular category of Soil Taxonomy represented by the term Halaquept is
   (a) Suborder (b) Great group
   (c) Subgroup (d) Family

586. With increase in pH, the extent of isomorphous substitution that contributes to the negative charge in silicate clays
   (a) increases (b) decreases
   (c) may increase or decrease (d) remains unchanged

587. Sodium Adsorption Ratio (SAR) of an irrigation water source is 22.5. The irrigation water is rated as
   (a) S1 Class. Low sodium hazard
   (b) S2 Class. Medium sodium hazard
   (c) S3 Class. High sodium hazard
   (d) S4 Class. Very high sodium hazard

588. The sequence of the soil Orders arranged according to their age (newest first)
   (a) Inceptisol, Oxisol, Entisol, Vertisol
   (b) Entisol, Ultisol, Aridisol, Oxisol
   (c) Entisol, Mollisol, Alfisol, Ultisol
   (d) Entisol, Spodosol, Ultisol, Oxisol

589. One gram equivalent H_2SO_4 equals to
   (a) 1 g H_2SO_4 (b) 2 g H_2SO_4
   (c) 49 g H_2SO_4 (d) 98 g H_2SO_4

590. The instrument used for measuring the plasticity of soils is
   (a) Pressure Plate apparatus (b) Pressure Membrane apparatus
   (c) Pycnometer (d) Atterberg apparatus

591. Principal organic cementing agents that help formation of soil aggregates are
   (a) lignins (b) proteins
   (c) polysaccharides (d) waxes

592. In Soil Taxonomy, the bottommost category for classifying soils is
   (a) Family (b) Series (c) Subgroup (d) Suborder

593. The feature of a colour denoted by the term 'chroma' is
   (a) dominant spectrum (b) intensity or brightness
   (c) purity (d) wave number

594. The microorganism which contribute the maximum to the biomass of soils are
   (a) Bacteria (b) Fungi
   (c) Actinomycetes (d) Protozoa

595. Downward movement of water from the surface to the lower depths of the soil is called
   (a) permeation (b) percolation
   (c) infiltration (d) runoff

596. According to the norms of the Soil Health Card Scheme of the Government of India, the grid area for drawing soil samples in the rainfed areas is
   (a) 1 ha (b) 2.5 ha (c) 10 ha (d) 25 ha

597. In Soil Taxonomy, a diagnostic subsurface horizon that is loamy textured, weakly cemented and is brittle, is referred by the term
   (a) Duripan (b) Fragipan (c) Plaggen (d) Sombric

598. With increase in pH, ionization of the functional groups in humus that contributes to the negative charge in the organic colloid
   (a) decreases (b) increases
   (c) may increase or decrease (d) remains unaffected

599. In Indian fertilizer market, the brand GROMOR is a In Indian fertilizer market, the brand GROMOR is a
   (a) complex fertilizer
   (b) mixed fertilizer
   (c) biofertilizer
   (d) micronutrient fertilizer

600. With increase in pH, the cation exchange capacity of smectite clays
   (a) increases (b) decreases
   (c) may increase or decrease (d) remains unchanged

601. In spectrophotometers, monochromatic radiation is selected by the use of a

(a) prism (b) grating
(c) either of the two (d) neither of the two

602. A fertilizer that develops alkalinity in soils is

(a) calcium sulphate (b) anhydrous ammonia
(c) ammonia solution (d) calcium cyanamide

603. Soils or parent material deposits referred to as lacustrines are formed inside

(a) rivers (b) sea and gulfs
(c) lakes (d) oceans

604. Movement of water along the soil surface is called

(a) permeation (b) percolation
(c) infiltration (d) runoff

605. In soils, the colloidal mass is accounted by

(a) layer lattice silicate, and iron and aluminium oxide clays
(b) allophane
(c) humus
(d) sum of the three

606. With increase in the alkalinity of soils, the cation exchange capacity of kaolinite

(a) increases (b) decreases
(c) may increase or decrease (d) remains unchanged

607. Currently, the number of essential plant nutrients for higher plants satisfying Arnon's criteria is

(a) 16 (b) 17 (c) 18 (d) 20

608. An expression for the Universal Soil Loss Equation (USLE) is

(a) $S = f(Cl, pm, b, r, t)$ (b) $S = f(b, r, t)$
(c) $A = f(RKLSCP)$ (d) $A = f(IKCLV)$

609. Nitrogen content of thiourea is

(a) 25% (b) 36.8% (c) 44% (d) 46%

610. The instrument used for measuring soil aggregation is

(a) Yoder's apparatus (b) Cone penetrometer
(c) Pycnometer (d) Atterberg apparatus

611. The 'Law of Minimum' in plant nutrition was proposed by
    (a) Justus von Liebig (b) Theodore de Saussure
    (c) Jean Baptiste Boussingault (d) Hans Jenny

612. *Rhizobia* bacteria responsible for biological nitrogen fixation are
    (a) heterotrophic anaerobic bacteria
    (b) heterotrophic aerobic bacteria
    (c) autotrophic anaerobic bacteria
    (d) autotrophic aerobic bacteria

613. In GPS, receivers used are
    (a) electronic clocks (b) quartz clocks
    (c) atomic clocks (d) mechanical clocks

614. Downy mildew of pearl millet is suppressed by
    (a) chloride (b) nickel (c) arsenic (d) chromium

615. Heat flow consequent to differences in the density of the molecules between two points of a body with unequal temperature is associated with
    (a) dissipation (b) conduction
    (c) convection (d) radiation

616. The mobile application launched by the Government of India in collaboration with Indian Meteorological Department and Indian Council of Agricultural Research for helping farmers in weather-based farming is named as
    (a) Krishi Kisan (b) Kisan Madad
    (c) eKisan (d) Meghdoot

617. According to Stokes' law, with increase in the density of the homogenous viscous medium, the terminal velocity of the falling particle (assuming other factors to remain unchanged)
    (a) increases (b) decreases
    (c) may increase or decrease (d) no relationship between the two

618. Light minerals are those minerals the density of which is lower than
    (a) 1.30 g/cm^3 (b) 1.76 g/cm^3
    (c) 2.30 g/cm^3 (d) 2.85 g/cm^3

619. The cation exchange capacity of kaolinite is low in comparison to that of montmorillonite because of
    (a) low external surface
    (b) low internal surface
    (c) low isomorphous substitution
    (d) all the three

620. The particular category of Soil Taxonomy represented by the whole term 'Grossarenic Argiaquoll, Fine, Silty, Illitic, Semiactive, Mesic, Rudali, *Eroded'* is

(a) Subgroup (b) Family
(c) Series (d) Phase

621. With increase in pH, the negative charge in silicate clays originating from the broken edges

(a) increases (b) decreases
(c) may increase or decrease (d) remains unchanged

622. Purple sulfur bacteria are

(a) chemoautotrophs (b) photoautotrophs
(c) aerobic heterotrophs (d) anaerobic heterotrophs

623. This soil material is an inorganic amorphous colloid

(a) kaolin (b) smectite (c) allophane (d) humus

624. Soil moisture characteristic curves depict the relationship between soil moisture and

(a) soil structure (b) soil texture
(c) soil minerology (d) soil moisture tension

625. Per cent recovery of applied phosphatic fertilizers is in the range of

(a) 20 to 40 (b) 40 to 60 (c) 60 to 70 (d) 70 to 80

626. A soil property that does not change much over a short period of time is

(a) texture (b) structure
(c) bulk density (d) water holding capacity

627. Magnesium is a component of

(a) plastocyanin (b) chlorophyll
(c) leghemoglobin (d) porphyrin

628. Albite is a

(a) potash feldspar (b) lime feldspar
(c) soda feldspar (d) magnesium feldspar

629. Soils rich in this clay mineral are unlikely to suffer from the deficiency of potassium

(a) montmorillonite (b) vermiculite
(c) illite (d) kaolinite

630. In an ideal soil, the per cent of the soil volume occupied by mineral matter is

(a) 20 to 30 (b) 45 to 55 (c) 60 to 70 (d) 70 to 80

631. The total surface area of the clay minerals kaolinite, illite, montmorillonite and chlorite is in the following order
   (a) montmorillonite > chlorite > illite > kaolinite
   (b) chlorite > illite > montmorillonite > kaolinite
   (c) chlorite > montmorillonite > illite > kaolinite
   (d) kaolinite > illite > montmorillonite > chlorite

632. All other factors remaining equal, the power of the monovalent cations to replace the cations on the soil exchange complexes is in the order:
   (a) $K^+ > Na^+ > Li^+ > H^+$ (b) $Na^+ > K^+ > H^+ > Li^+$
   (c) $H^+ > K^+ > Na^+ > Li^+$ (d) $K^+ > H^+ > Na^+ > Li$

633. The expression $\Sigma\, [Ca^{2+}] + [Mg^{2+}] + [K^+] + [Na^+] =$ CEC (all in m.e./100 g soil) is almost invariably true for
   (a) acid soils (b) calcareous soils
   (c) acid sulphate soils (d) high rainfall area soils

634. This soil material is an organic amorphous colloid
   (a) kaolin (b) smectite (c) allophane (d) humus

635. Per cent recovery of applied fertilizers is the least in the case of
   (a) nitrogenous fertilizers (b) phosphatic fertilizers
   (c) potassic fertilizers (d) sulphur fertilizers

636. Manufacture of ammonium sulphate fertilizer by Gypsum process involves the following chemical reaction
   (a) decomposition (b) double decomposition
   (c) hydration (d) hydrolysis

637. This property is not a characteristic of inorganic soil colloids
   (a) high plasticity (b) high surface area
   (c) adsorbed cations (d) high refractive index

638. The unit for expressing the bulk density of soils is
   (a) g/cm^2 (b) Mg/m^2 (c) dg/m^2 (d) Mg/m^3

639. Diatomoceous earth is used in fertilizers and fertilizer mixtures as a
   (a) carrier (b) filler
   (c) separator (d) coating agent

640. Deaminated bonemeal is produced by the removal of this constituent from raw bonemeal
   (a) carbohydrates (b) proteins
   (c) lipids (d) heteropolysaccharides

641. In Soil Taxonomy, sodium rich subsurface horizon is referred as
(a) Cambic (b) Argillic (c) Natric (d) Sodic

642. In highly alkaline soils, loss of fertilizer nitrogen takes place mostly in the form of
(a) N_2 (b) NH_3 (c) NO (d) NO_2

643. In Soil Taxonomy, the category represented by the term Typic Haplorthod, Sandy, Siliceous, Frigid is
(a) Great group (b) Subgroup
(c) Family (d) Series

644. Size of the particles falling in a viscous medium must be large enough to prevent them from Brownian movement is a factor to be considered in testing
(a) Darcy's law (b) Fourier's law
(c) Fick's law (d) Stokes' law

645. Microorganisms which are adapted for survival at temperatures above 45^0 C are called
(a) mesophiles (b) thermophiles
(c) psychrophiles (d) cryophiles

646. A consequential effect of the global warming of the planet earth is
(a) unpredictable climate change
(b) rise in sea level causing inundation of coastal areas
(c) melting of the earth's glaciers
(d) all the three

647. The foul smell often produced due to putrefactive fermentation during the process of making composts from urban wastes are due to the formation of
(a) thiophosphates (b) sulphides
(c) mercaptans (d) thiosulphate

648. Soils or parent material deposits formed by glaciers as the agency of transportation are called
(a) outwash (b) till (c) fluvial (d) glacial

649. This is not an accepted term for describing the consistence of wet soils
(a) slightly sticky (b) sticky
(c) very sticky (d) very highly sticky

650. According to Fick's law of diffusion, the molar flux of a gas across a plane due to diffusion is proportional to
(a) volume of the gas
(b) concentration gradient of the gas
(c) purity of the gas
(d) specific heat of the gas

651. Fourier's law is related to the flow of
(a) liquids (b) gases (c) fluids (d) heat

652. The cation exchange capacity of kaolinite is low in comparison to that of vermiculite because of
(a) coarser particle size
(b) lower exposed surface
(c) lower isomorphous substitution
(d) all the three

653. If the bulk density of a soil is increased from 1.4 to 2.1 g/cm^3, the porosity of the soil is (particle density of the soil is 2.8 g/cm^3)
(a) decreased to its one fourth (b) increased by its one fourth
(c) decreased to its half (d) increased by its half

654. Microorganisms that grow well in acidic pH are
(a) bacteria (b) fungi
(c) protozoa (d) actinomycetes

655. This is not an accepted term for describing the consistence of wet soils
(a) slightly plastic (b) plastic
(c) very plastic (d) very highly plastic

656. High total surface area of the clay mineral montmorillonite is due to
(a) fine particle size (b) high internal surface
(c) both the two (d) neither of the two

657. The ratio of organic matter to nitrogen in soil organic matter is around
(a) 20:1 (b) 5:1 (c) 50:1 (d) 100:1

658. The notation 'h' in Planck's equation '$E = h\frac{c}{\lambda}$' denotes
(a) velocity of light (b) Planck's constant
(c) wave number (d) wavelength

659. N-Serve, a nitrification inhibitor, is a substituted
(a) pyridine (b) pyrimidine (c) triazine (d) triazole

660. This is not an Atterberg's consistence limit
(a) Plastic limit (b) Viscous limit
(c) Liquid limit (d) Shrinkage limit

661. The cation exchange capacity of the clay minerals kaolinite, illite, montmorillonite and vermiculite follows the following order
(a) montmorillonite < vermiculite < kaolinite < illite
(b) vermiculite < illite < montmorillonite < illite
(c) vermiculite < montmorillonite < illite < kaolinite
(d) kaolinite < illite < montmorillonite < vermiculite

662. A reducing agent frequently used in soil analysis laboratory is
(a) $SnCl_2$ (b) H_2O_2
(c) sodium dithionite (d) EDTA

663. The internal surface of the clay mineral montmorillonite is
(a) nil (b) very low (c) low (d) high

664. The Genetic system of Soil Classification classifying soils into Zonal, Intrazonal and Azonal classes was proposed by
(a) Marbut (b) Dokuchaev (c) Smith D Guy (d) Joffe

665. Low surface area of the clay mineral kaolinite is due to
(a) coarse particle size (b) low internal surface
(c) both the two (d) none of the two

666. Velocity of light in vacuum is
(a) 3×10^8 cm/s (b) 3×10^8 m/s
(c) 3×10^8 dm/s (d) 3×10^8 km/s;econd

667. Unit for measuring radioactivity of radioactive substances is
(a) Curie (b) Rontgen (c) Becquerel (d) all the three

668. The practice of growing erosion permitting and erosion resisting crops in alternate strips across the slope as a measure to check soil erosion by water in steep slopes is called
(a) contour strip cropping (b) field strip cropping
(c) water strip cropping (d) gully strip cropping

669. 'Feel method' with a small amount of soil sample taken on the palm is performed to have an approximate idea of the soil's
(a) structure (b) texture
(c) consistence (d) strength

670. Parent materials that lie over the underlying bedrock from which they have been formed are referred to as parent materials of this kind
(a) transported (b) colluvial
(c) sedentary (d) transformed

671. DTPA is used as a/an
   (a) synthetic mulch
   (b) antitranspirant
   (c) micronutrient fertilizer
   (d) extractant for available micronutrients

672. This nanotubular aluminosilicate clay mineral is found to occur in the soils belonging to the order Andisol
   (a) nontronite (b) imogolite (c) sauconite (d) saponite

673. The equation relating the properties of a soil to be a function of the five soil forming factors was first suggested by
   (a) Joffe (b) Jenny (c) Liebig (d) Dokuchaev

674. The epipedon that has a thick black horizon at or near, but within 30 cm of the soil surface, is
   (a) Grossarenic (b) Plaggen
   (c) Melanic (d) Folistic

675. The quantity of heat required to raise the temperature of a unit mass of soil by 1^0 C is the soil's
   (a) specific heat (b) heat capacity
   (c) thermal conductivity (d) thermal diffusivity

676. India's first-ever coal gasification based fertilizer plant is located at
   (a) Phulpur in Uttar Pradesh (b) Kalol in Gujarat
   (c) Talcher in Odisha (d) Trombay in Maharashtra

677. The benefits of green manuring are
   (a) improvement of soil physical properties
   (b) addition of all plant nutrients to soil
   (c) prevention soil erosion
   (d) all the three

678. In Land Capability Classification, the Class that serves the purpose for use only for wildlife and watershed is
   (a) Class I to Class IV (b) Class V to Class VIII
   (c) Class VI, VII and VII (d) Class VIII

679. Man-made epipedon high in phosphorus is
   (a) Mollic (b) Anthropic (c) Umbric (d) Histic

680. One of the followings is not a benefit rendered by green manuring
   (a) addition of plant nutrients to soil
   (b) improvement of soil physical environment
   (c) change of unfavourable soil texture
   (d) prevention of soil erosion

681. The albedo of sandy soils in comparison to that of clayey soils is
(a) lower (b) higher
(c) same (d) no relationship between the two

682. The word manure is derived from the word 'manoeuvrer' which is a
(a) French word (b) Chinese word
(c) Greek word (d) Latin word

683. An insecticide that has been banned primarily for its high persistence in soil is
(a) aldrin (b) BHC (c) DDT (d) all the three

684. Quartzite, a metamorphic rock, is produced from the following rock
(a) soapstone (b) sandstone
(c) conglomerate (d) shale

685. Micron, a unit of length, is expressed by the notation
(a) μ (b) nμ (c) mm (d) μdm

686. The term Versene is used to refer to
(a) EDTA (b) DTPA (c) EDDHA (d) NTA

687. Schoonover described a simple method for the laboratory determination of this property of soils
(a) Cation Exchange Capacity
(b) Exchangeable Sodium Per centage
(c) Lime Requirement of acid soils
(d) Gypsum Requirement of alkali soils

688. Colorimeter used for measuring the intensity of the radiation transmitted by a solution in experiments in analytical chemistry laboratory is based upon
(a) absorption spectroscopy (b) emission spectroscopy
(c) IR spectroscopy (d) UV spectroscopy

689. Planck's equation that relates the energy of radiation to wavelength is
(a) $E = h \nu / \lambda$ (b) $E = h c\lambda$
(c) $E = h c/\lambda$ (d) $E = h c$

690. The micronutrients manganese, iron, nickel, copper and zinc are all *d* block elements of the following series in the Periodic Table of the elements
(a) 4*d* (b) 3*d* (c) 5*d* (d) 6*d*

691. The chemical composition of the FCO approved molybdenum fertilizer in India is
(a) $CaMoO_4$ (b) MoO_3
(c) $(NH_4)_2MoO_4$ (d) $(NH_4)_6Mo_7O_{24}.4\ H_2O$

692. In local parlance, the term *Bhangar* refers to
(a) new alluvium (b) old alluvium
(c) both the two (d) neither of the two

693. It is convenient to express the wavelength of electromagnetic radiation in ultraviolet region in terms of
(a) kilometre (b) millimetre
(c) decimetre (d) nanometer

694. If the concentration of OH^- ion of a solution is 10^{-3} mole/litre, the pH of the solution is
(a) 3 (b) 11 (c) – 3 (d) – 11

695. The combustible gas produced in biogas plants is
(a) methane (b) ethane (c) propane (d) butane

696. 'Hidden hunger' of plant nutrients refers to a situation where
(a) visual deficiency symptoms appear when the nutrient is deficient
(b) visual deficiency symptoms do not appear even though the nutrient is deficient
(c) the nutrient is not deficient and yet deficiency symptoms appear
(d) the nutrient is available and yet deficiency symptoms appear

697. The instrument lysimeter is used to measure
(a) wind velocity
(b) evaporation water loss
(c) evapotranspiration water loss
(d) density of viscous fluids

698. The dissolution of this substance in water produces considerable heat (exothermic)
(a) concentrated H_2SO_4 (b) solid NaOH
(c) solid $CaCl_2$ (d) all the three

699. If the pOH of a solution is 4, pH of the same is
(a) – 4 (b) 4 (c) – 10 (d) 10

700. Fertile soil deposits rich in silt and clay formed near the mouth of rivers are called
(a) alluvial terrace (b) outwash
(c) delta (d) flood plain

701. *Khaira* disease in rice is caused due to the deficiency of
(a) iron (b) zinc (c) copper (d) manganese

702. The method of soil moisture determination by using aluminium boxes and oven is
   (a) gravimetric method (b) gypsum block method
   (c) neutron scattering method (d) tensiometer method

703. The method of fertilizer application that shows the maximum Fertilizer Use Efficiency (FUE) is
   (a) drip fertigation (b) top dressing
   (c) band placement (c) split application

704. The height of capillary rise of water in soil is inversely proportional to the
   (a) purity of water
   (b) radius of the tube
   (c) hydrostatic pressure difference
   (d) surface tension of water

705. A nonexpanding silicate clay is
   (a) kaolinite (b) illite
   (c) montmorillonite (d) vermiculite

706. As soil pH rises, the P constituent of single super phosphate is eventually converted to
   (a) triple super phosphate (b) monocalcium phosphate
   (c) dicalcium phosphate (d) tricalcium phosphate

707. Diagnostic subsurface horizon that is highly acidic ($pH < 3.5$) is called
   (a) Hydrochloric (b) Sulphuric
   (c) Nitric (d) Acetic

708. The term 'Steppe' in describing soils of cultivated areas refers to an ecoregion of
   (a) plain grassland (b) tropical forest lands
   (c) coniferous forest lands (d) hydromorphic lands

709. The plant nutrient that often exhibits luxury consumption is
   (a) nitrogen (b) phosphorus (c) potassium (d) sulphur

710. The soil that has the highest bulk density is
   (a) sandy (b) silty (c) clayey (d) loamy

711. This step is not performed in mechanical analysis of soils
   (a) flocculation (b) dispersion
   (c) sedimentation (d) fractionation

712. Flame photometer, the analytical instrument in use for determining potassium and sodium contents of samples, operates within this region of the electromagnetic spectrum
(a) visible (b) ultraviolet (c) near infrared (d) far infrared

713. The dissolution of this substance in water produces an ionic solution
(a) urea (b) glucose (c) $MgSO_4$ (d) all the three

714. The problems associated with hidden hunger of plant nutrients can be unmistakably handled by resorting to
(a) soil testing (b) plant analysis
(c) soil fertility evaluation (d) all the three

715. Oxidation of soil organic matter by H_2O_2 is a step performed in the experiment on
(a) soil structure evaluation (b) soil moisture determination
(c) soil mechanical analysis (d) soil infiltration study

716. Soil scientist regarded as the 'Father of Pedology' is
(a) Hans Jenny (b) Justus von Liebig
(c) Thomas Way (d) V. V. Dokuchaev

717. This is not one among the twelve Orders of Soil Taxonomy
(a) Lithosol (b) Spodosol (c) Gelisol (d) Andisol

718. This metal is a key component in the chemical structure of plastocyanin
(a) iron (b) copper
(c) manganese (d) magnesium

719. Monochromatic electromagnetic radiation of wavelength of 650 nm falls within this region of the electromagnetic spectrum
(a) ultraviolet (b) visible
(c) near infrared (d) far infrared

720. In particle size analysis of soils, sodium hexametaphosphate is added as a/an
(a) aggregating agent (b) flocculating agent
(c) oxidizing agent (d) dispersing agent

721. Nebular hypothesis to explain the theory for the formation of the planet earth was proposed by
(a) Kuiper (b) Laplace
(c) Hoyles (d) Chamberlin and Moulton

722. Fried and Dean first suggested the concept of
   (a) DRIS concept of plant nutrition
   (b) Inverse Ratio law
   (c) Q/I relationship
   (d) 'A' value for the dynamics of drawing a plant nutrient from two sources

723. One of the following rocks is not a monomineralitic rock
   (a) quartzite (b) sandstone (c) granite (d) limestone

724. Soil horizon that shows an accumulation of silica, sesquioxides and clay due to illuviation is
   (a) A horizon (b) B horizon
   (c) C horizon (d) D horizon

725. In Soil Taxonomy, the Order for a soil under the Subgroup Lithic Fibristel is
   (a) Histosol (b) Andisol (c) Gelisol (d) Entisol

726. The metamorphic rock produced from granite, an igneous rock, is
   (a) gneiss (b) schist (c) slate (d) quartzite

727. A bulky organic manure is
   (a) neem cake (b) fish meal
   (c) guano (d) urban compost

728. An endopedon common in soils is
   (a) Umbric (b) Histic (c) Argillic (d) Anthropic

729. Flame photometry is based upon
   (a) absorption spectrophotometry
   (b) emission spectrophotometry
   (c) infra red spectrophotometry
   (d) ultra violet spectrophotometry

730. The cation of this metal is intimately associated with the functioning of the energy rich compound ATP
   (a) magnesium (b) zinc
   (c) iron (d) copper

731. The diameter of soil separates is usually expressed in terms of
   (a) micron (b) milli micron
   (c) milli meter (d) centi meter

732. Grass tetany disease in cattle is caused due to the consumption of forages with low content of
   (a) magnesium (b) sulphur
   (c) iron (d) zinc

733. Sulphur is absorbed by plants in the form of
   (a) S (b) S^{2-} (c) SO_3^{2-} (d) SO_4^{2-}

734. For the capillary rise of water in soil, the relationship between the height and radius is
   (a) direct (b) inverse
   (c) direct or inverse (d) no relationship between the two

735. The symbionts that fix atmospheric nitrogen in *Azolla pinnata* are
   (a) algae (b) bacteria (c) fungi (d) ferns

736. The metallic element present in vitamin B12 as a structural constituent is
   (a) iron (b) copper
   (c) cobalt (d) molybdenum

737. An early soil scientist regarded as the 'Father of Field Plot Techniques' is
   (a) Justus von Liebig (b) Theodore de Saussure
   (c) Hens Jenny (d) Jean Baptiste Boussingault

738. Liming of acid soils increases the availability of
   (a) iron (b) copper
   (c) manganese (d) molybdenum

739. Laboratory analysis revealed that 1 m^2 area of an acid soil required 90 g calcium for liming. With limestone as the liming material, the Lime Requirement of the soil in tons/ha is (purity of the liming material is 90%)
   (a) 1.5 (b) 2.25 (c) 2.5 (d) 4.5

740. The availability of this nutrient remains relatively less affected by soil pH
   (a) nitrogen (b) potassium
   (c) sulphur (d) calcium

741. A gas in the atmosphere that contributes to acid rains is
   (a) H_2S (b) CH_4 (c) NH_3 (d) SO_2

742. Ultraviolet radiation reaching the earth from the sun is intercepted by
   (a) atmosphere (b) ionosphere
   (c) lithosphere (d) ozonosphere

743. Walkley and Black developed a simple method for the determination of soil
(a) organic carbon (b) total nitrogen
(c) sulphur (d) available nitrogen

744. 'Grey speck' disease in some plants like oat and barley is caused due to the deficiency of
(a) iron (b) copper (c) zinc (d) manganese

745. Average per cent CO_2 content (v/v) of the upper soil layers is around
(a) 0.03 (b) 0.10 (c) 0.25 (d) 1.2

746. The expression for deriving the Residual Sodium Carbonate (RSC) of irrigation water is
(the brackets indicate concentrations of the ions in m. e. per litre)
(a) $RSC = [CO_3^{2-}] - ([Ca^{2+}] + [Mg^{2+}])$
(b) $RSC = ([CO_3^{2-}] + [HCO_3^-]) - ([Ca^{2+}] + [Mg^{2+}])$
(c) $RSC = ([CO_3^{2-}] + [HCO_3^-]) - [Ca^{2+}]$
(d) $RSC = ([CO_3^{2-}] + [HCO_3^-]) - [Mg^{2+}]$

747. The process of increase of the concentration of pollutants through the food chain is called
(a) bioremediation (b) biomagnification
(c) biotoxication (d) biotrophication

748. A common nonfossil fuel is
(a) coal (b) petroleum (c) hydrogen (d) natural gas

749. The process of the conversion of sediments to sedimentary rocks is called
(a) conglomeration (b) cementation
(c) compaction (d) lithification

750. Soil horizon that shows depletion of silica, sesquioxides and clay due to eluviation is
(a) A horizon (b) B horizon (c) C horizon (d) D horizon

751. The most abundant mineral in sandstone rock is
(a) feldspars (b) quartz
(c) micas (d) muscovite

752. The equilibrium constant K_{eq} of the reaction AB = A + B is
(a) $K_{eq} = \frac{[A]\,[B]}{[AB]}$ (b) $K_{eq} = \frac{[A]\,[B]}{[B]}$
(c) $K_{eq} = \frac{[A]\,[B]}{[A]}$ (d) $K_{eq} = \frac{[AB]}{[A]\,[B]}$

753. To obey Stokes' law, the particles falling in the viscous medium must be
(a) angular (b) rectangular
(c) prismatic (d) spherical

754. In high pH alkaline soils, soluble phosphorus is fixed as insoluble
(a) monocalcium phosphate (b) dicalcium phosphate
(c) tricalcium phosphate (d) all the three

755. Hornblende is a mineral of the following class
(a) amphibole (b) mica
(c) olivine (d) feldspar

756. An example of a feldspar mineral is
(a) muscovite (b) forsterite
(c) hornblende (d) orthoclase

757. According to the Soil Health Card Scheme of the Government of India, a soil is considered 'medium' with regard to available nitrogen, if its content is
(a) 140 to 280 kg/ha (b) 280 to 560 kg/ha
(c) 560 to 700 kg/ha (d) more than 700 kg/ha

758. Gypsum Requirement analysis in the laboratory revealed that 1 m^2 area of an alkali soil required 80 g calcium. The Gypsum Requirement (molecular weight 172) of the soil in tons/ha is (purity of gypsum is 95%)
(a) 1.81 (b) 3.62 (c) 4.25 (d) 4.5

759. In soil testing laboratory, Conductivity Bridge instrument is used to measure
(a) total soluble salt content of the soil
(b) total insoluble salt content of the soil
(c) total salt content of the soil
(d) precipitated salt content of the soil

760. Basaltic rocks on weathering form
(a) alluvial soils (b) brown soils
(c) black soils (d) sandy soils

761. Acid igneous rocks on weathering form
(a) red soils (b) clayey soils
(c) black soils (d) sandy soils

762. Core sampler method is used for measuring soil
(a) bulk density (b) particle density
(c) hydraulic conductivity (d) infiltration capacity

763. This mineral nutrient is essential for nitrogen fixing microbes in legumes
(a) silicon (b) nickel (c) cobalt (d) vanadium

764. With increase in clay content, the bulk density of a soil
(a) decreases (b) increases
(c) may increase or decrease (d) remains unaffected

765. Diagnostic surface horizon (epipedon) that is thick, dark coloured, and does possess high base saturation and strong structure is
(a) Mollic (b) Umbric (c) Ochric (d) Histic

766. A soil property commonly associated with soil crusting is the soil's
(a) high organic matter content (b) low silt content
(c) high sodium content (d) high gypsum content

767. COLE of a soil is indicative of the soil's
(a) particle density (b) resistance to percolation
(c) rate of infiltration (d) swelling capacity

768. Soil crusts of different types such as slaking, infiltrating, coalescing and sieving crusts are soil crusts of the following kind
(a) structural (b) erosional
(c) depositional (d) cryptogamic

769. The clay mineral that causes cracks in Vertisol soils upon drying is
(a) kaolinite (b) illite
(c) montmorillonite (d) chlorite

770. A gas in the atmosphere that contributes to acid rains is
(a) H_2S (b) CH_4 (c) NH_3 (d) NO_2

771. The letter G in the standard abbreviation GPS stands for
(a) geographical (b) geological
(c) geostationary (d) global

772. Permeability of soils is expressed in terms of
(a) millimeters per hour (b) litres per hour
(c) litres per centimeter (d) litres per centimeter per hour

773. Soil crusting causes an increase in soil's
(a) bulk density (b) particle density
(c) cation exchange capacity (d) pH

774. c-Axis spacings of clay minerals are expressed in terms of
(a) cm (b) mm (c) m (d) Å

775. The instrument used to measure the depth of water table is
(a) sonometer (b) piezometer
(c) hydrometer (d) penetrometer

776. Dapog method of raising rice nursery was introduced by
(a) Indonesia (b) South Korea
(c) Malayasia (d) Philippines

777. Terracing is an effective method of checking soil erosion in
(a) desert area (b) hilly area
(c) plain area (d) low rainfall area

778. This radioisotope does have a half-life of 5730 years
(a) H^3 (b) Mo^{99} (c) C^{14} (d) Fe^{59}

779. Renewal of soil air by its exchange with the atmospheric air takes place by the process/es of
(a) mass flow
(b) diffusion
(c) both mass flow and diffusion
(d) neither mass flow nor diffusion

780. In Land Capability Classification, the lands under Class VIII are shown in maps in this colour
(a) red (b) orange (c) purple (d) green

781. Soil crusting causes deterioration in soil's
(a) texture (b) structure
(c) cation exchange capacity (d) pH

782. The plant nutrient the essentiality of which was established by Sommer and Lipman is
(a) iron (b) manganese
(c) zinc (d) molybdenum

783. In Land Capability Classification, the Land Capability Subclass is further divided into a number of
(a) Land Capability Sub-subclasses
(b) Land Capability Units
(c) Land Capability Families
(d) Land Capability Phases

784. The specific surface of soil separates follow the order:
   (a) fine sand > coarse sand > silt > clay
   (b) coarse sand > silt > fine sand > clay
   (c) clay > fine sand > silt > coarse sand
   (d) clay > silt > fine sand > coarse sand

785. Amount of nitrate (NO_3^-) required for replacing 2 milliequivalents of (not as 'SO_4^{2-}) (sulphate) by anion exchange process is
   (a) 2 mg (b) 4 mg (c) 124 mg (d) 248 mg

786. A fertilizer that adds neither acidity nor alkalinity to soil is
   (a) urea (b) urea solution
   (c) ammonium nitrate (d) calcium ammonium nitrate

787. A crop that prefers nitrogen in ammonium form for considerable part of its growth is
   (a) rice (b) wheat (c) maize (d) pea

788. Diagnostic subsurface horizon (endopedon) that is characterized by accumulation of salts is
   (a) Gypsic (b) Salic (c) Calcic (d) Petrocalcic

789. Specific characters of the master horizons of soils are indicated by adding this kind of notation as suffixes after the abbreviations of the master horizons (A, B etc.)
   (a) numerals such as 1,2 etc. e.g. A1, A2, A3
   (b) lower case letters such as a,b, c etc. e.g. Ap, B_t
   (c) numerals such as 1, 2 after the lower case letter e.g. A_p1, B_t1
   (d) Roman numbers such as I, II etc. e.g. AI, BII

790. Drinking water high in nitrate nitrogen causes
   (a) anemia (b) methemoglobinemia
   (c) improper bone formation (d) night blindness

791. Bulk of the gaseous exchange in soil system takes place through
   (a) mass flow (b) diffusion
   (c) convection (d) contact exchange

792. For the restoration of highly degraded soils, plantation of the land is recommended with
   (a) grasses (b) tall trees
   (c) seedlings of trees (d) orchard crops

793. Decomposition of this constituent in organic matter delays its rate of decomposition in soil
(a) oligosaccharides (b) polysaccharides
(c) proteins (d) lignins

794. The instrument used for assessing consumptive use of water by plants is
(a) anemometer (b) pycnometer
(c) lysimeter (d) piezometer

795. *Bacillus megatherium* is a
(a) nitrogen fixing bacteria
(b) nitrifying bacteria
(c) denitrifying bacteria
(d) phosphate solubilizing bacteria

796. This epipedon is not common in India
(a) Mollic (b) Histic (c) Umbric (d) Anthropic

797. In Land Capability Classification, the Classes that serve the purpose only for grazing
(a) Class I to Class IV (b) Class I to Class V
(c) Class V to Class VII (d) Class V and Class VIII

798. For the determination of a soil chemical property, say soil organic carbon content, the mass of a soil sample of 1 g drawn from a large mass of composite soil sample represents the following fraction of the mass of the hectare furrow slice soil (assume the soil bulk density and depth of furrow slice to be 1.335 g/cm^3 and 15 cm, respectively)
(a) two millionth (b) half a billionth
(c) half a millionth (d) two billionth

799. The most important chemical process involved in the weathering of rocks and minerals is
(a) oxidation (b) reduction
(c) hydration (d) hydrolysis

800. In Soil Taxonomy, one of the followings is not a Suborder under the Order Inceptisol
(a) Gelept (b) Ustept (c) Udept (d) Oxept

801. The gas produced in the puddled rice soil is
(a) NO (b) SO_2 (c) CH_4 (d) NO_2

802. The per cent nitrogen content of diammonium phosphate is
(a) 11 (b) 18 (c) 26 (d) 46

803. Thermal diffusivity is the ratio of thermal conductivity to
(a) heat capacity (b) thermal retentivity
(c) thermal resistivity (d) heat flux

804. FCO stipulated the minimum per cent (w/w) water soluble P_2O_5 content of diammonium phosphate is
(a) 16 (b) 46 (c) 41 (d) 52

805. An expression for soil heat flux is
(a) $\frac{dT}{dz}$ (b) $\frac{dz}{dT}$ (c) $K\frac{dT}{dz}$ (d) $K\frac{dz}{dT}$

806. Chamberlin and Moulton's proposed theory/hypothesis to explain the formation of the planet earth is known as
(a) Nebular hypothesis (b) Big bang theory
(c) Planetesimal hypothesis (d) Supernova hypothesis

807. The process of intermixing of soil horizons by external forces is called
(a) pedoturbation (b) gleization
(c) physicochemical weathering (d) melanization

808. Soil heat flux is upward during
(a) morning (b) noon
(c) midday (d) midnight

809. In accepted hydrological terms, waterlogged areas are the land areas where the water table is
(a) within 2 m depth from soil surface
(b) between 2 to 4 m depth from soil surface
(c) between 4 to 5 m depth from soil surface
(d) within 5 m depth from soil surface

810. According to WHO specifications, the concentration of nitrate in drinking water must not exceed
(a) 5 mg/L (b) 10 mg/L (c) 20 mg/L (d) 50 mg/L

811. Biofertilizers in essence are
(a) bulky organic manures (b) green maures
(c) organic manures (d) cultures of microorganisms

812. Man made sod like horizon created by years of manuring is
(a) Mollic (b) Anthropic
(c) Plaggen (d) Histic

813. The per cent nitrogen content of monoammonium phosphate is
(a) 11 (b) 18 (c) 26 (d) 46

814. The symbol K in the expression $Q = c\ K\ A\ \frac{h_2 - h_1}{L}$ of Darcy's law denotes
(a) flux density (b) flow velocity
(c) hydraulic gradient (d) hydraulic conductivity

815. Nutrient element that imparts drought resistance to plants is
(a) nitrogen (b) phosphorus
(c) potassium (d) magnesium

816. An effective *in situ* method for soil moisture conservation is
(a) drip irrigation (b) sprinkler irrigation
(c) fertigation (d) mulching

817. Mg/m^3 is the unit for expressing this property of soils is
(a) bulk density (b) particle density
(c) both the two (d) neither of the two

818. Diagnostic surface horizon (epipedon) that is light coloured, contains low organic matter and may be hard and massive when dry is
(a) Mollic (b) Umbric (c) Ochric (d) Histic

819. The one among the four that has the highest albedo is
(a) land area covered by vegetation
(b) bare land without vegetation
(c) water surface
(d) ice

820. A temperature reading of 25^0 C equals to
(a) 45^0 F (b) 77^0 F (c) 95^0 F (d) 105^0 F

821. One of the followings is not associated with podzolization soil forming process
(a) movement of iron and aluminium oxides from surface to lower soil layers
(b) movement of silica from surface to lower soil layers
(c) acid soil pH
(d) cool humid climate

822. Water holding capacity is the maximum in
(a) sandy soils (b) sandy loam soils
(c) loamy soils (d) clayey soils

823. The contribution by the different processes in wind erosion is in the following order
(a) saltation > surface creep > suspension
(b) saltation > suspension > surface creep
(c) suspension > saltation > surface creep
(d) surface creep > saltation > suspension

824. Allophane, an amorphous inorganic colloid, is more abundant in soils of
(a) humid regions
(b) volcanic origin
(c) intense weathering
(d) alluvial deposits

825. Black soils usually come under the soil Order
(a) Entisol
(b) Inceptisol
(c) Histosol
(d) Vertisol

826. Most recommended of the soil:water ratios for preparing the soil suspension for measuring the pH of soils is
(a) 1:1 (b) 1:2.5 (c) 1:10 (d) 1:100

827. The per cent total P_2O_5 content of monoammonium phosphate is
(a) 16 (b) 42 (c) 46 (d) 52

828. Energy of an electromagnetic radiation is directly related to its
(a) frequency
(b) wavelength
(c) both the two
(d) neither of the two

829. An example of a sedimentary rock is
(a) syenite
(b) diorite
(c) shale
(d) slate

830. A temperature of 105^0 F is equal to
(a) 37.5^0 C
(b) 40.5^0 C
(c) 45.5^0 C
(d) 50.5^0 C

831. A physical process associated with heat loss from soil is
(a) infiltration
(b) translocation
(c) evaporation
(d) transpiration

832. In laboratory studies on micronutrients, dithizone is used as a/an
(a) chelate
(b) extractant
(c) chromogenic agent
(d) buffer

833. Murexide is an indicator of the following type
(a) acid base indicator
(b) metal ion indicator
(c) precipitation indicator
(d) adsorption indicator

834. Disodium EDTA solution is used as a titrant in the determination of
(a) calcium
(b) magnesium
(c) both calcium and magnesium
(d) neither calcium nor magnesium

835. Wilting is the most general deficiency symptom of
(a) boron (b) molybdenum
(c) manganese (d) chlorine

836. Matric potential in soil is due to
(a) capillary rise of soil water
(b) cohesive action among soil particles
(c) adhesive action among soil particles
(d) adsorption due to soil colloids

837. An environment friendly pesticide should have
(a) selective toxicity to the target pest
(b) short persistence in the environment
(c) rapid detoxification in case of ingestion in animals
(d) all the three

838. Canopy Temperature Depression is inversely proportional to
(a) canopy temperature (b) soil temperature
(c) air temperature (d) soil moisture

839. In space, heat is transferred in the form of
(a) thermal energy (b) electromagnetic radiation
(c) kinetic energy (d) potential energy

840. The bulk density, particle density and porosity of a soil are 0.7 g/cm^3, 2.8 g/cm^3, and 75%, respectively. If the porosity of the soil is now reduced to 50%, the bulk density of the soil is
(a) decreased to its half (b) increased by its half
(c) doubled (d) increased by its one fourth

841. Conversion of nitrate nitrogen to oxides of nitrogen and/or elemental nitrogen as done soil is a process of
(a) hydrolysis (b) oxidation
(c) reduction (d) hydration

842. In the expression of Stokes' law, the viscosity of the fluid is usually represented by the Greek alphabet
(a) δ (b) ε (c) η (d) θ

843. The ionic radii of the following monovalent cations follows the sequence
(a) $Na^+ > K^+ > H^+ > Li^+$ (b) $Li^+ > Na^+ > H^+ > K^+$
(c) $K^+ > Na^+ > Li^+ > H^+$ (d) $H^+ > K^+ > Na^+ > Li^+$

844. The solubility products (K_{sp}) of the sparingly soluble salts Ca $(OH)_2$, $CaCO_3$ and $CaSO_4$ are 5.5 x 10^{-6}, 4.8 x 10^{-9} and 1.2 x 10^{-6}, respectively. Accordingly, the solubility of the salts (g/L) is in the order
(a) $Ca(OH)_2 < CaCO_3 < CaSO_4$ (b) $CaCO_3 < CaSO_4 < Ca(OH)_2$
(c) $CaCO_3 < Ca(OH)_2 < CaSO_4$ (d) $Ca(OH)_2 < CaSO_4 < CaCO_3$

845. An example of a metamorphic rock is
(a) syenite (b) diorite (c) shale (d) slate

846. The changes in spectral reflectivity/emissivity with time are called
(a) spectral variations (b) temporal variations
(c) spatial variations (d) optical variations

847. With increase in soil acidity, anion exchange capacity (AEC) and cation exchange capacity (CEC) are altered as follows
(a) AEC decreases, but CEC increases
(b) AEC increases, but CEC decreases
(c) both AEC and CEC decrease
(d) both AEC and CEC increase

848. The equivalent weight of oxalic acid ($H_2C_2O_4$) in the reaction '$H_2C_2O_4 \rightarrow 2\ CO_2 + 2\ H$' is
(a) 90 (b) 180 (c) 120 (d) 45

849. The indicator used in $K_2Cr_2O_7$ vs $FeSO_4$ titration in acid medium is
(a) diphenylamine (b) mixed indicator
(c) phenolphthalein (d) methyl red

850. In comparison to black soils, the red soils have
(a) lower CEC and lower per cent base saturation
(b) lower CEC and higher per cent base saturation
(c) higher CEC and lower per cent base saturation
(d) higher CEC and higher per cent base saturation

851. One of the following properties is not a function of state
(a) enthalpy (b) entropy
(c) internal energy (d) concentration

852. If in the experiment on the determination of chlorides, the titre value is 5.2 ml 0.02 N $AgNO_3$ for 10 ml soil water extract, then g per litre soluble chloride content of the soil water extract is

(a) 0.00104 (b) 0.00208
(c) 0.03688 (d) 0.36878

853. Lignins of soil organic matter are

(a) phenolic polymers (b) polycarboxylic acids
(c) heteropolysaccharides (d) quinones

854. Sulphate content of soils using $BaCl_2$ is determined by

(a) colorimetry (b) flame photometry
(c) titrimetry (d) turbidimetry

855. An approximate estimate of TDS (mg/L) of a water source derived from its Electrical Conductivity (dS/m) is obtained from the following empirical relationship

(a) TDS = EC x 10 (b) TDS = EC x 100
(c) TDS = EC x 64 (d) TDS = EC x 640

856. If the bulk density of a soil is reduced from 1.4 to 0.7 g/cm^3, the porosity of the soil is changed as (particle density of the soil is 2.8 g/cm^3)

(a) decreased by its 25% (b) increased by its 25%
(c) decreased by its 50% (d) increased by its 50%

857. In Soil Taxonomy, the soils are classified under a particular Order on the basis of their

(a) parent materials (b) Soil Moisture Regimes
(c) Soil Temperature Regimes (d) Diagnostic horizons

858. Elemental oxygen content of soil air is

(a) less than 21% (b) more than 25%
(c) more than 30% (d) less than 5%

859. Rothamsted Experimental Station in UK is famous for

(a) long term field studies in soil and agricultural research
(b) research on Soil Survey and Classification
(c) research in soil clay minerology
(d) research in soil geology

860. A light weight low specific gravity and frothed rock is

(a) shale (b) slate (c) asbestos (d) pumice

861. In soil forming processes, lime, silica and iron oxides are all
(a) flocculating agents (b) dispersing agents
(c) cementing agents (d) oxidizing agents

862. According to Poiseuille's law, the rate of flow of a liquid through a narrow tube is inversely proportional to the
(a) radius of the tube (b) pressure gradient
(c) temperature (d) viscosity of the liquid

863. Prairie soils which account for a significant area of the world's ice-free land are classified under the soil Order
(a) Vertisol (b) Mollisol (c) Aridisol (d) Ultisol

864. One picometer is equal to
(a) 10^{-15} meter (b) 10^{-12} meter
(c) 10^{-10} meter (d) 10^{-9} meter

865. Thermal conductivity of soils is expressed in terms of
(a) cal/s/ ^{0}C (b) cal/cm/s/^{0}C (c) cal/cm/s (d) cal/g/s/ ^{0}C

866. Diffusion of a fluid in a medium is a function of
(a) concentration gradient
(b) diffusion coefficient of the medium
(c) cross sectional area
(d) all the three

867. Diagnostic surface horizon (epipedon) that is high in organic matter and wet during some part of the year is
(a) Mollic (b) Umbric (c) Ochric (d) Histic

868. *Streptomyces* genera are
(a) bacteria (b) fungi
(c) actinomycetes (d) protozoa

869. Conversion of simple inorganic nitrogen like ammonium compounds to complex organic nitrogen of organic matter as done in the case of immobilization is a process of
(a) hydrolysis (b) oxidation
(c) reduction (d) hydration

870. In nuclear decay, the radioisotope C^{14} is an emitter of
(a) alfa particles with high ionization power
(b) hard beta particles with high penetrating power
(c) soft beta particles with very low penetrating power
(d) very short wavelength electromagnetic radiation

871. One of the followings is not a FCO approved zinc fertilizer in India
   (a) zinc sulphate monohydrate ($ZnSO_4.H_2O$)
   (b) zinc sulphate heptahydrate ($ZnSO_4.7\ H_2O$)
   (c) zinc chloride anhydrous ($ZnCl_2$)
   (d) zinc-EDTA (Zn-EDTA)

872. The bacteria involved in the process of nitrification in soils are
   (a) heterotrophs (b) saprophytes
   (c) photoautotrophs (d) chemoautotrophs

873. The redox potential of chemical equilibra is determined by using
   (a) Bragg's equation (b) Freundlich equation
   (c) Planck's equation (d) Nernst equation

874. The Sodium Adsorption Ratio (SAR) of irrigation water is derived from the following expression
   (the cations are expressed in terms of m.e./L water)
   (a) $\frac{[Na^+]}{\sqrt{\frac{[Ca^{2+}]+[Mg^{2+}]}{2}}}$ (b) $\frac{[Na^+]}{\frac{[Ca^{2+}]+[Mg^{2+}]}{2}}$
   (c) $\frac{[Na^+]}{\sqrt{\frac{[Ca^{2+}]+[Mg^{2+}]}{4}}}$ (d) $\frac{[Na^+]}{[Ca^{2+}]+[Mg^{2+}]}$

875. Equivalent weight of $Fe_2(SO_4)_3$ (molecular weight is 399.9) in its reduction to $FeSO_4$ is
   (a) 399.88 (b) 133.29 (c) 199.94 (d) 99.97

876. The volume of water to be added for raising the pH of 100 ml solution from 6 to 7 is
   (a) 100 ml (b) 200 ml (c) 900 ml (d) 1000 ml

877. In soil classification, tundra soils are classified as
   (a) Histosol (b) Cryosol
   (c) Regosol (d) Lithosol

878. A crop highly tolerant to soil acidity is
   (a) alfalfa (b) sugar beet
   (c) peanut (d) citrus

879. This endopedon is light coloured
   (a) Albic (b) Molic (c) Placic (d) Melanic

880. In Soil Taxonomy, dark coloured base rich (>50%) soils of grasslands are classified under the Order
   (a) Mollisol (b) Inceptisol
   (c) Entisol (d) Alfisol

881. Diagnostic subsurface horizon (epipedon) that is characterized by accumulation and cementation of insoluble $CaSO_4$ wherein the fragments do not slake in water is
(a) Gypsic (b) Petrogypsic (c) Calcic (d) Petrocalcic

882. In waterlogged soils, reduction of the compounds of this element is the most conspicuous chemical reaction
(a) phosphorus (b) magnesium
(c) potassium (d) iron

883. Volume (V) of a cylindrical core soil sampler is given by the formula
(a) $\pi d^2 h$ (b) $\pi r^2 h$ (c) $\frac{\pi r^2 h}{4}$ (d) $\frac{\pi r^2 h}{2}$

884. According to Darcy's law, the flow of a liquid through a porous medium like soil is inversely proportional to
(a) hydraulic conductivity
(b) hydraulic gradient
(c) cross sectional area of the porous medium
(d) length over which the hydraulic drop is taking place

885. The letter G in the standard abbreviation GIS stands for
(a) geographic (b) geologic
(c) geostationary (d) global

886. Thermal diffusivity of soils is expressed in terms of
(a) cal/cm/s (b) cal/g/s/^{0}C (c) cm^2/s (d) s/cm^2

887. Capital letters like A, B etc. are used to designate soils'
(a) master horizons
(b) deviations within master horizons
(c) diagnostic surface horizons
(d) diagnostic subsurface horizons

888. Ferrous ammonium sulphate is known as
(a) Mohr's salt (b) Rochelle salt
(c) Epsom salt (d) Glauber's salt

889. The ease with which water passes through a bulk mass of soil is the soil's
(a) infiltration (b) percolation
(c) conductivity (d) permeability

890. Energy of electromagnetic radiation is inversely related to the
(a) frequency of the radiation (b) wavelength of the radiation
(c) both the two (d) neither of the two

891. One of the following processes does not take place in puddled rice field
(a) conversion of $MnCl_2$ to $MnCl_4$
(b) formation of CH_4 from the decomposition of organic matter
(c) formation of sulphide from $(NH_4)_2SO_4$
(d) increase in the activity of denitrifying bacteria

892. Given that λ, C_v, and c represent the thermal conductivity thermal conductivity, volumetric heat capacity and specific heat respectively, of a soil, the thermal diffusivity (α) of the soil is expressed as
(a) $\alpha = \frac{\lambda}{C_V}$ (b) $\alpha = \frac{C_V}{\lambda}$ (c) $\alpha = \frac{C_V c}{\lambda}$ (d) C_v c

893. For nomenclature in Soil Taxonomy, the formative element for the soil Order Gelisol is
(a) *ol* (b) *oll* (c) *el* (d) *eg*

894. An expression for calculating the Per cent Base Saturation (PBS) of a soil is (CEC and the cations are expressed in terms of m.e./100 g soil)
(a) $\frac{[Ca^{2+}] + [Mg^{2+}] + [K^{+}]}{CEC}$ x 100
(b) $\frac{[Ca^{2+}] + [Mg^{2+}] + [Na^{+}]}{CEC}$ x 100
(c) $\frac{[Ca^{2+}] + [K^{+}] + [Na^{+}]}{CEC}$ x 100
(d) $\frac{[Ca^{2+}] + [Mg^{2+}] + [K^{+}] + [Na^{+}]}{CEC}$ x 100

895. According to the Soil Health Card Scheme of the Government of India, a soil is considered 'high' with regard to available potassium, if its content is
(a) 60 to 120 kg/ha (b) 120 to 280 kg/ha
(c) more than 280 kg/ha (d) more than 200 kg/ha

896. If the Great Group of a soil is Salustert, then according to Soil Taxonomy, Ustert designates the soil's
(a) Suborder (b) Subgroup (c) Family (d) Series

897. The unit for expressing the particle density of soils is
(a) g/cm^2 (b) Mg/m^3
(c) dg/m^2 (d) both (a) and (b)

898. The chemical formula for Glauber's salt is
(a) $Na_2CO_3.10\ H_2O$ (b) $CaSO_4.2\ H_2O$
(c) $Na_2SO_4.10\ H_2O$ (d) $MgSO_4.7\ H_2O$

899. The micronutrient cation that has fungicidal value is
(a) manganese (b) copper
(c) cobalt (d) boron

900. Reduction in particle size with no change in chemical composition is a characteristic of
(a) physical weathering (b) chemical weathering
(c) physicochemical weathering (d) biological weathering

901. Saltation and surface creep are processes associated with
(a) wind erosion (b) water erosion
(c) surface runoff (d) leaching

902. c-Axis spacing of kaolinite, illite and montmorillonite is in the following order:
(a) illite > kaolinite > montmorillonite
(b) kaolinite > illite > montmorillonite
(c) illite > montmorillonite > kaolinite
(d) montmorillonite > illite> kaolinite

903. A farmer has to apply single super phosphate (SSP) @ 300 kg/ha. Assuming that SSP is not available and his field size is 0.5 ha, the amount of triple super phosphate (TSP) he requires is (P_2O_5 contents of SSP and TSP are 16 and 46%, respectively)
(a) 24 kg (b) 52 kg (c) 150 kg (d) 225 kg

904. Regarding optimum temperature required for growth, majority soil organisms are
(a) psychrophiles (b) mesophiles
(c) thermophiles (d) all the three

905. This mineral contains both iron and titanium
(a) stengite (b) ilmenite
(c) magnetite (d) vivianite

906. Red soils are widespread in this state of India
(a) West Bengal (b) Andhra Pradesh
(c) Arunachal Pradesh (d) Tamil Nadu

907. If the Lime Requirement of an acid soil is 5.5 tons/ha and due to nonavailability, burnt lime (CaO) is to be used instead of limestone, then the quantity of burnt lime for a field plot of 0.5 ha is
(a) 1.54 tons (b) 3.08 tons
(c) 2.75 tons (d) 4.91 tons

908. One of the followings is NOT a unit of radioactivity
(a) Curie (Ci) (b) Becquerel (Bq)
(c) Dalton (D) (d) Rutherford (Rd)

909. This substance is used as a primary standard base for the standardization of prepared acid solutions
(a) NaOH (b) KOH (c) borax (d) $Ca(OH)_2$

910. This mineral is black in colour
(a) muscovite (b) biotite
(c) haematite (d) malachite

911. According to FCO Specifications of Fertilizations, nitrophosphate (20-20-0) must contain ammonical and nitrate nitrogen in
(a) 10% each (b) entirely in ammonical form
(c) entirely in nitrate form (d) 15% and 5%

912. Cobalt is a constituent of
(a) hemoglobin (b) leghaemoglobin
(c) plastocyanin (d) vitamin B12

913. Biosuper is an intimate mixture of rock phosphate and
(a) elemental sulphur (b) gypsum
(c) pyrites (d) sulphate of potash

914. Eutrophication in water bodies pertains to the bloom in the population of
(a) higher plants (b) aquatic ferns
(c) fungi (d) algae

915. This substance is used as a primary standard acid for the standardization of prepared alkali solutions
(a) HCl (b) potassium hydrogen phthalate
(c) H_2SO_4 (d) $NaHCO_3$

916. Isobutylidenediurea is a
(a) coated fertilizer (b) slow release fertilizer
(c) nitrification inhibitor (d) nitrification promoter

917. Flame Ionization Detector (FID) is a component of the instrumentation of
(a) Flame photometer
(b) atomic absorption spectrophotometer
(c) Gas Liquid Chromatograph (GLC)
(d) absorption spectrophotometer

918. Organic matter with wide C:N ratio should not be applied to a field with a standing crop because of the problem associated with
(a) denitrification (b) ammonia volatilization
(c) mineralization (d) immobilization

919. Soils suitable for growing rubber plantation are
(a) black soils (b) red soils
(c) laterite soils (d) alluvial soils

920. Slickensides are often noticed in
(a) soils rich in humus
(b) soils rich in sand
(c) soils rich in nonexpanding clays with little swelling and shrinkage
(d) soils rich in expanding clays with high swelling and shrinkage

921. In potentiometric determinations such as the measurement of H^+ concentration or pH of a solution, the electrode that records a constant potential is called the
(a) reference electrode (b) indicator electrode
(c) glass electrode (d) none of the three

922. Biological weathering involves the processes of
(a) physical weathering (b) chemical weathering
(c) both the two (d) neither of the two

923. A gas contributing to greenhouse effect is
(a) CO_2 (b) CH_4
(c) both the two (d) neither of the two

924. Preindependent India's first fertilizer plant established at Ranipet in Tamil Nadu manufactured
(a) ammonium sulphate (b) urea
(c) ammonium nitrate (d) single super phosphate

925. A substance added to put off the offensive odour due to putrefactive fermentation during the process of making composts from urban wastes is
(a) chlorine (b) sodium thiosulphate
(c) copper sulphate (d) bleaching powder

926. A carrier material for *Rhizobium* culture is
(a) peat (b) lignite (c) charcoal (d) all the three

927. The base of the equilateral triangle of the soil textural diagram represents the per cent content of
(a) clay (b) silt (c) sand (d) fine sand

928. Rich deposits of gypsum are found in this state of India
(a) Haryana (b) Odisha
(c) Sikkim (d) Rajasthan

929. Red colour of red soils is due to the presence
   (a) oxides of iron (b) oxides of aluminium
   (c) muscovite (d) biotite

930. The concentration of exchangeable H^+, K^+, Na^+, Ca^{2+} and Mg^{2+} ions of a soil are 3, 39, 69, 150 and 18 mg/100 g soil, respectively. Assuming the soil not to have any other exchangeable cation, the soil meets the ESP requirement of a/an
   (a) saline soil (b) acid soil
   (c) calcareous soil (d) alkali soils

931. The metamorphic rock produced from sandstone, a sedimentary rock, is
   (a) gneiss (b) schist (c) slate (d) quartzite

932. A soil on analysis was found to contain 1.035% organic matter. Considering the average C:N of agricultural soils to be 10:1 to 15:1, the per cent nitrogen content of the soil is
   (a) 0.400 to 0.600 (b) 0.040 to 0.060
   (c) 0.600 to 0.900 (d) 0.060 to 0.090

933. An effective practice to check soil loss by wind erosion is
   (a) contour bunding (b) terrace cropping
   (c) strip cropping (d) ridge and furrow cropping

934. In particle size analysis of soils, deflocculating agent added to maintain the stability of the dispersed soil system is
   (a) sodium dithionite (b) sodium hexametaphosphate
   (c) DTPA (d) hydrogen peroxide

935. The particular category of Soil Taxonomy represented by the term Orthents is
   (a) Suborder (b) Great group (c) Subgroup (d) Family

936. According to the Soil Health Card Scheme of the Government of India, a soil is considered 'high' with regard to organic carbon, if its content is
   (a) 0.25 to 0.50 % (b) 0.50 to 0.75 %
   (c) more than 0.75 % (d) more than 0.50 %

937. Calomel electrode used in potentiometers consists of a pool of this metal in contact with a paste of its chloride salt
   (a) mercury (b) silver
   (c) antimony (d) none of the three

938. The expression for deriving Residual Sodium Bicarbonate (RSBC) of irrigation water is

(the brackets indicate concentrations of the ions in m. e. per litre)

(a) $[HCO_3^-] - ([Ca^{2+}] + [Mg^{2+}])$
(b) $([CO_3^{2-}] + [HCO_3^-]) - ([Ca^{2+}] + [Mg^{2+}])$
(c) $[HCO_3^-] - [Ca^{2+}]$
(d) $[HCO_3^-] - [Mg^{2+}]$

939. Amounts of ammonium sulphate and single super phosphate to be applied in place of 150 kg DAP for its nonavailability are (% N content of ammonium sulphate is 20.6; % N and P_2O_5 contents of DAP are 18 and 46; % P_2O_5 content of SSP is 16)

(a) 131 kg A.S. + 131 kg SSP (b) 131 kg A.S. + 231 kg SSP
(c) 131 kg A.S. + 431 kg SSP (d) 181 kg A.S. + 431 kg SSP

940. This epipedon is so named because of its light colour

(a) Ochric (b) Mollic (c) Folistic (d) Placic

941. In remote sensing, the device that gathers electromagnetic radiation, converts it into a signal and presents it in a form suitable for obtaining information about the object under investigation, is called

(a) lens (b) sensor (c) scanner (d) camera

942. Swamps, bogs, marshes, and estuaries are examples of

(a) watersheds (b) waterlogged soils
(c) wetlands (d) submerged soils

943. Leaching Requirement (LR) for the reclamation of saline soils is derived by the expression

(a) $LR = \frac{EC_{iw}}{EC_{dw}}$ (b) $LR = \frac{EC_{dw}}{EC_{iw}}$
(c) $LR = \frac{EC_{iw} \text{ x } EC_{dw}}{EC_{iw}}$ (d) $LR = \frac{EC_{iw} \text{ x } EC_{dw}}{EC_{dw}}$

(EC_{iw} and EC_{dw} indicate the electrical conductivity of the irrigation and drainage water, respectively)

944. The analytical instrument that uses a hollow cathode lamp as the source of radiation is

(a) flame photometer
(b) UV visible spectrophotometer
(c) atomic emission spectrophotometer
(d) atomic absorption spectrophotometer

945. Crop logging is a method of
   (a) soil fertility evaluation
   (b) plant analysis for assessing requirements of nutrients for crop production
   (c) phytotoxicity of excessive use of fertilizers
   (d) testing suitability of fertilizers

946. A crop quite tolerant to high salt concentration in soil is
   (a) beans (b) pea (c) grape (d) sugar beet

947. In Soil Taxonomy, highly weathered soils are classified under the Order
   (a) Spodosol (b) Oxisol
   (c) Histosol (d) Mollisol

948. A solution of muriate of potash tests
   (a) acidic (b) neutral
   (c) slightly alkaline (d) highly alkaline

949. Ramsar sites refer to
   (a) sites of degraded soils (b) designated wastelands
   (c) designated wetlands (d) specific areas of polluted soils

950. Echard, Hollard and Chresard are terms for use in describing
   (a) soil water plant relationship
   (b) soil thermal regime plant relationship
   (c) soil chemical behaviour plant relationship
   (d) soil redox behaviour plant relationship

951. Amount of sulphate of potash equivalent to 150 kg of muriate of potash with regard to K_2O content is (K_2O contents of potash and muriate of potash are 50 and 60%)
   (a) 90 kg (b) 120 kg (c) 180 kg (d) 240 kg

952. P^{31} and P^{32} are
   (a) isomers of each other (b) isomorphs of each other
   (c) isotopes of each other (d) isobars of each other

953. Inderite is a mineral of magnesium and
   (a) molybdenum (b) boron
   (c) copper (d) manganese

954. Amount of $CaCO_3$ required for the neutralization of acidity produced by the application of 180 kg ammonium sulphate is (acid equivalent of ammonium sulphate is 110)
   (a) 180 kg (b) 198 kg (c) 220 kg (d) 262 kg

955. Molybdenum blue method is used for the colorimetric/ spectrophotometric determination of
(a) phosphate (b) nitrate
(c) sulphate (d) ammonium

956. In plant analysis, organic matter of plant samples is destroyed by
(a) dry ashing (b) digestion by triacid mixture
(c) either of the two (d) neither of the two

957. Plant nutrient that increases the oil content of groundnut and other oilseed crop plants is
(a) nitrogen (b) phosphorus
(c) potassium (d) sulphur

958. Soils or parent material deposits referred as dunes are formed by the following agency of transportation
(a) water (b) wind (c) ice (d) gravity

959. The total water present in soil is known as Hollard. The total water available to plants is known as
(a) Echard (b) Hollard (c) Chresard (d) Conard

960. The constituent of organic matter most resistant to microbial degradation is
(a) cellulose (b) hemicelluloses
(c) waxes (d) lignins

961. Critical Soil Test Level Concept of nutrient management in plants was suggested by
(a) Beaufils (b) Fried and Dean
(c) Cate and Nelson (d) Sommer and Lipmann

962. A crop very sensitive to soil acidity is
(a) grapes (b) sugar beet
(c) tobacco (d) citrus

963. The commercially available form of this fertilizer does have tints of various colours due to the presence of impurities
(a) muriate of potash (b) sulphate of potash
(c) ammonium sulphate (d) urea

964. If the Electrical Conductivity (EC) of the saturation paste of a soil is 2 dS/m at 25^0 C, the same in terms of μmho/cm is
(a) 0.2 (b) 2 (c) 20 (d) 2000

965. A commonly used carrier gas for sweeping the substances to be separated in Gas Liquid Chromatography is
(a) oxygen (b) methane
(c) acetylene (d) nitrogen

966. Soapstone used as a coating agent in the manufacture of fertilizers is primarily made of
(a) serpentine (b) pyrophyllite
(c) talc (d) hornblende

967. A solution of ammonium sulphate tests
(a) acidic (b) neutral
(c) slightly alkaline (d) highly alkaline

968. Soil amendment used for the reclamation of alkali soils is
(a) limestone (b) dolomite
(c) gypsum (d) basic slag

969. In particle size analysis of soils, the chemical added to the soil sample for the destruction of soil organic matter is
(a) sodium dithionite (b) sodium hexametaphosphate
(c) DTPA (d) hydrogen peroxide

970. According to FCO, 1985 Specifications amended in 2021, the total nitrogen (as N) per cent by weight, minimum of vermicompost to be commercially available in market is
(a) 1 (b) 2 (c) 3 (d) 5

971. This fertilizer does bear the risk of explosive hazard due to which its use is curtailed
(a) ammonium nitrate (b) calcium ammonium nitrate
(c) ammonium chloride (d) monoammonium phosphate

972. The primary objective of converting the crystalline powdery forms of fertilizers to their granular forms is to
(a) increase their water solubility
(b) decrease their contact area with the soil
(c) decrease their shelf-life
(d) increase their chemical reactivity

973. In Ostwald's process, nitric acid is produced by the catalytic oxidation of
(a) N_2 (b) NH_3
(c) NH_2NH_2 (d) NH_2OH

974. Amount of $CaCO_3$ equivalent following the application of 175 kg calcium cyanamide fertilizer is (base equivalent of the fertilizer is 63)
(a) 63 kg (b) 110 kg (c) 126 kg (d) 175 kg

975. Colemanite is a mineral of
(a) boron (b) molybdenum
(c) zinc (d) manganese

976. Recommended depth of soil sampling for advisory purposes like fertilizer recommendation, lime requirement etc. is usually
(a) 6 inches (b) 1 foot (c) 3 feet (d) 6 feet

977. In soil sampling, the collected soil samples after recommended drying is ground and passed through a
(a) 0.6 mm sieve (b) 1 mm sieve
(c) 2 mm sieve (d) 4 mm sieve

978. A triacid mixture common in use for the digestion of plant samples is nitric, sulphuric and
(a) acetic acid (b) perchloric acid
(c) dichloroacetic acid (d) trichloroacetic acid

979. Manufacturing facility of this fertilizer does have a prilling tower
(a) calcium ammonium nitrate (b) urea
(c) urea ammonium phosphate (d) ammonium polyphosphate

980. The source of radiation for the analytical instrument atomic absorption spectrophotometer is a/an
(a) ordinary tungsten lamp (b) hydrogen lamp
(c) deuterium lamp (d) hollow cathode lamp

981. Deficiency of calcium causes bitter pit disease in the fruits of
(a) pears (b) apple (c) orange (d) banana

982. Normality (N) of commercial concentrated sulphuric acid is around
(a) 1 (b) 10 (c) 35 (d) 50

983. The disadvantage of ammonium nitrate for its use as a fertilizer are its high
(a) explosive hazard (c) high hygroscopicity
(b) high acidity (d) all the three

984. For titrimetric analysis in the laboratory, EDTA is used as a
(a) monosodium salt (b) disodium salt
(c) monocalcium salt (d) dicalcium salt

985. The atomic numbers of the plant nutrient elements K, Ca, Mg and S follows the sequence
(a) S > Ca > Mg > K (b) Ca > K > S > Mg
(c) Ca > Mg > S > K (d) K > S > Mg > Ca

986. Pahala blight in sugarcane is caused due to the deficiency of
(a) zinc (b) copper (c) manganese (d) boron

987. For the determination of available potassium in soils, a representative mass of the soil is treated with
(a) 1 N ammonium acetate, pH 7.0
(b) 0.5 M $NaHCO_3$, pH 8.5
(c) 1% ammonium citrate
(d) 0.01 M $CaCl_2$

988. A fertilizer that contains 25% total nitrogen is
(a) ammonium nitrate (b) calcium ammonium nitrate
(c) monoammonium phosphate (d) diammonium phosphate

989. Cuvettes (sample holders) of spectrophotometers for working in the visible range of the electromagnetic spectrum are made of
(a) glass (b) porcelain (c) plastic (d) ceramic

990. The fertilizer that can be recommended for correcting the deficiency of sulphur is
(a) ammonium chloride (b) single super phosphate
(c) triple super phosphate (d) dolomite

991. The term 'pomace' in agricultural chemistry refers to a manure obtained from
(a) carcasses of animals (b) residues of oil extracting plants
(c) night soil (d) kitchen wastes

992. One among the following fertilizers is commercially available in the form of super granules
(a) urea (b) calcium ammonium nitrate
(c) single super phosphate (d) triple super phosphate

993. Deficiency symptoms of the plant nutrients that are mobile in plant systems are noticed first in
(a) older leaves (b) newer leaves
(c) both older and newer leaves (d) apical leaves

994. Devarda's alloy used in the determination of total nitrogen content of soils by Kjeldahl method is made up of aluminium, copper and
(a) chromium (b) mercury (c) iron (d) zinc

995. A crop quite sensitive to high salt concentration in soil is
(a) pulses (b) date palm
(c) rice (d) sugar beet

996. A fertilizer that is commercially available in the form of prills is
(a) calcium ammonium nitrate (b) urea
(c) ammonium sulphate (d) sulphate of potash

997. Ramsar Convention recommendations that became effective in December 1975 pertains to
(a) land degradation (b) soil erosion
(c) habitat destruction (d) wetlands

998. Atomiser for causing a mist of the sample solution is a component of this analytical instrument in
(a) colorimeter (b) spectrophotometer
(c) flame photometer (d) conductivity bridge

999. The most widely used source of hydrogen for the manufacture of nitrogenous fertilizers in India is
(a) water (b) coal (c) natural gas (d) naphtha

1000. The particular category of Soil Taxonomy represented by the term Vertic Ustorthent is
(a) Suborder (b) Great group (c) Subgroup (d) Family

1001. The usual oxidizing agent for determining and expressing the Chemical Oxygen Demand (COD) for assessing the pollution load of water sample is
(a) $KMnO_4$ (b) $K_2Cr_2O_7$
(c) H_2O_2 (d) concentrated HNO_3

1002. This soil property is a universal character of soil health and quality
(a) soil pH (b) soil structure
(c) Soil Organic Carbon (SOC) (d) soil moisture retention capacity

1003. This mineral is a trioctahedral mica
(a) halloysite (b) muscovite
(c) chlorite (d) biotite

1004. The two ions which together contribute to the acidity of the soil solution are
(a) Mg^{2+} and Al^{3+} (b) H^+ and Ba^{2+}
(c) H^+ and Al^{3+} (d) Ba^{2+} and Al^3

1005. In India, this fertilizer alone accounts for more than 50% of the total consumption of all commercial chemical fertilizers
   (a) calcium ammonium nitrate
   (b) urea
   (c) diammonium phosphate (DAP)
   (d) NPK complexes

1006. Daniel Hillel's much referred book is on this branch of Soil Science
   (a) Soil Fertility (b) Soil Physics
   (c) Soil Chemistry (d) Soil Classification

1007. Nitrogenase, the enzyme responsible for biological nitrogen fixation, is an enzyme of the following Class
   (a) Oxidoreducatase (b) Transferase
   (c) Ligase (d) Lyase

1008. The water solubility of monocalcium phosphate (MCP), dicalcium phosphate (DCP) and tricalcium phosphate (TCP) is in the following order
   (a) TCP > DCP > MCP (b) DCP > MCP > TCP
   (c) DCP > TCP > MCP (d) MCP > DCP > TCP

1009. The fertilizer calcium cyanamide ($CaCN_2$) is hydrolyzed in soil to form
   (a) CaO and HCN (b) CaO and CH_3NH_2
   (c) CaO and NH_2NH_2 (d) $CaCO_3$ and NH_3

1010. *Ambrosia artenaisiofolia* is used as an indicator plant for deficiency of this nutrient element
   (a) copper (b) zinc
   (c) manganese (d) molybdenum

1011. K_2O to K conversion factor is
   (a) 2.29 (b) 0.44 (c) 1.21 (d) 0.83

1012. In Land Capability Classification, the lands under Class I are shown in soil maps in this colour
   (a) yellow (b) orange (c) purple (d) green

1013. For manufacturing single super phosphate, the P_2O_5 content of rockphosphate must be a minimum of
   (a) 16% (b) 31% (c) 46% (d) 52%

1014. This soil Order has so far not been reported in India
   (a) Spodosol (b) Oxisol
   (c) Gelisol (d) All the three

1015. With regard to water of crystallization present in the molecules of borax, the fully hydrated form of the substance is
(a) monohydrate (b) dihydrate
(c) heptahydrate (d) decahydrate

1016. According to Soil Taxonomy, the number of soil Orders found in India is
(a) 8 (b) 9 (c) 10 (d) 12

1017. The analytical instrument used for the determination of micronutrient cations is
(a) emission spectrophotometer
(b) gas liquid chromatograph
(c) polarograph
(d) atomic absorption spetrophotometer

1018. The ionic form of phosphorus produced by the dissolution of single super phosphate in water is
(a) $(PO_4)^{3-}$ (b) $(HPO_4)^{2-}$ (c) $(H_2PO_4)^{-}$ (d) $(H_2P_2O_7)^{2-}$

1019. Nitrogen content of nano urea liquid currently available in Indian market is
(a) 4% (b) 23% (c) 44% (d) 46%

1020. An enzyme involved in the first of the two steps of nitrification executed by the *Nitrosomonas* group of bacteria is
(a) nitrate reductase
(b) ammonia monooxygenase
(c) nitrite oxidase
(d) nitrite reductase

1021. Isomorphous substitution that confers electrical charge to silicate clays refers to the substitution of one ion by another one with comparable
(a) mass (b) size (c) electrical charge (d) valency

1022. Particle size of urea in nano urea liquid formulation currently available in Indian market is
(a) less than 1 mm (b) 1 mm to 4 mm
(c) 20 to 50 μm (d) 20 to 50 nm

1023. In healthy plants, the N:S ratio should be between
(a) 80 to 100:1 (b) 50 to 60:1
(c) 7 to 9:1 (d) 2 to 3:1

1024. For Desulfovibrio bacteria, the terminal acceptor of electrons for its respiratory requirements in waterlogged soils is
   (a) sulphate (SO_4^{2-}) (b) sulphite (SO_3^{2-})
   (c) sulphide (S^{2-}) (d) tetrathionate ($S_4O_6^{2-}$)

1025. Presence of $Ca(NO_3)_2$ in a complex fertilizer or fertilizer mixture enhances the following characteristic of the product
   (a) drillability (b) flowability
   (c) hygroscopicity (d) shelflife

1026. In the experiment on the determination of Ca+Mg, 10 ml soil water extract was found to contain 0.087 milli equivalent of Ca+Mg together. Assuming 10 ml soil water extract prepared from 0.5 g soil to contain 1.2 mg Ca, the per cent Mg content of the soil is
   (a) 0.024 (b) 0.240 (c) 0.6480 (d) 0.0648

1027. Strongly acidic soil pH helps suppressing the incidence of this disease of potato
   (a) late blight (b) pink rot
   (c) scab (d) potato virus Y

1028. Characteristic Soil Moisture Regime (SMR) and Soil Temperature Regime (STR) of soils are mostly used in assigning the following category of Soil Taxonomy
   (a) Order (b) Suborder
   (c) Great Group (d) Subgroup

1029. Chile salt peter is a nitrate of
   (a) sodium (b) potassium
   (c) calcium (d) magnesium

1030. If the C:N and C:P ratios of a soil are 10:1 and 150:1, respectively, then the N:P ratio of the soil is
   (a) 10:1 (b) 15:1 (c) 100:1 (d) 150:1

1031. Chisel ploughing is resorted to manage and amend the following limitation of a soil
   (a) gleyed subsoil
   (b) subsurface hardpan
   (c) waterlogged condition
   (d) unfavourable crumb soil structure

1032. In False Colour Composite (FCC), the colour in which vegetation is shown is
   (a) green (b) blue (c) purple (d) red

1033. In making soil maps, the basic unit for mapping is
(a) Family (b) Phase (c) Series (d) Subgroup

1034. Biotite is a primary mineral under the following mineralogical class
(a) micas (b) feldspars
(c) amphiboles (d) pyroxenes

1035. In works in Soil Taxonomy and Soil Survey, a soil property that is determined very approximately in the field itself by a semiquantitative method and later in the laboratory by a recommended method is the soil's
(a) colour (b) texture
(c) structure (d) bulk density

1036. Coniferous vegetation is a characteristic of the soils belonging to the Order
(a) Aridisol (b) Mollisol
(c) Spodosol (d) Vertisol

1037. *Kankar*, a colloquial term often used in soil science, refers to concretions of
(a) lime carbonate (b) gypsum
(c) pyrites (d) fine silt

1038. The distance between the middle of the camera lens and the focal plane is called
(a) optical length (b) focal length
(c) spatial length (d) spectral length

1039. Among the four soil Orders viz. Alfisol, Mollisol, Oxisol and Vertisol, infertile soils are more likely in soils belonging to the Order
(a) Alfisol (b) Vertisol (c) Mollisol (d) Oxisol

1040. Extent of carbonation reaction in chemical weathering of rocks and minerals is more prominent in soils belonging to the Order
(a) Mollisol (b) Vertisol (c) Aridisol (d) Gelisol

1041. H bonding is typified by the attraction between a partially positive atom of hydrogen and a partially negative atom of
(a) oxygen (b) nitrogen
(c) fluorine (d) all the three

1042. Ligands are neutral or ionic species that bind to
(a) metals ions (b) clays
(c) humic substances (d) all the three

1043. The characteristic property used in identifying clay minerals is their
   (a) X ray diffractograms
   (b) Differential Thermal Analysis (DTA) curves
   (c) both X ray diffractograms and Differential Thermal Analysis (DTA) curves
   (d) Nuclear Magnetic Resonance (NMR) spectra

1044. Among the four clay minerals viz. chlorite, montmorillonite, vermiculite and kaolinite, the one that is the most chemically weathered is
   (a) chlorite (b) montmorillonite
   (c) vermiculite (d) kaolinite

1045. A substance that can accept as the terminal acceptor of elctrons for soil microbial respiration in the absence of molecular oxygen is
   (a) NO_3^- (b) SO_4^{2-} (c) Fe^{3+} (d) all the three

1046. The available nitrogen content of an agricultural soil is 500 mg/kg soil. Following soil application of 283 kg urea/ha (46% N), the per cent available nitrogen increases to (assume 1 ha furrow slice soil to weigh 2.2 million kg soil)
   (a) 0.0280 (b) 0.0559 (c) 0.0420 (d) 0.0955

1047. DTPA is an extractant of choice for the determination of available micronutrient cation contents of soils is for this property of the substance
   (a) it forms ionic bonds with the micronutrient cations
   (b) it forms hydrophobic bonds with the micronutrient cations
   (c) large molecular size
   (d) high number of dents

1048. A factor that determines the weatherability of minerals is their
   (a) order of crystallization (b) contents of iron and aluminium
   (c) increased basicity (d) all the three

1049. Soil mirorganisms that are filamentous are
   (a) fungi (b) actinomycetes
   (c) fungi and actinomycetes (d) fungi, actinomycetes and algae

1050. Formation of urea by heating ammonium cyanate, as done by Friedrich Wohler in his famous Urea Synthesis in the laboratory, is a chemical reaction of the following kind
   (a) oxidation (b) dehydration
   (c) isomerisation (d) carbonation

II. Filling the Blanks

Fill in the blanks with right answer/s

1. Soil formation process characterized by the removal of soil materials in suspension or solution such as silica, sesquixides, clay etc. from upper soil depths is referred to as ______________________.
2. A brucite sheet is a structural component of the layer lattice crystal structure of the silicate clay ______________________.
3. If the concentration of Ca^{2+}, Mg^{2+}, CO_3^{2-} and HCO_3^- ions of an irrigation water source are 80, 24, 165 and 61 mg/L, respectively, the Residual Sodium Carbonate (RSC) of the water source is _________.
4. If the Great Group of a soil is Salustert, then according to Soil Taxonomy, the category of the soil denoted by Ustert is ________________.
5. Two important variables of climate influencing soil formation process are ______________ and ________________.
6. In silicon tetrahedron, one Si ion remains equidistant from four _________ ions.
7. The unit for expressing the Oxygen Diffusion Rate (ODR) requirement of crops is ________________.
8. The nutrient the deficiency of which causes the *khaira* disease in rice is ____________________.
9. The solid sphere of the earth is referred to as ____________________.
10. Sulphate content is multiplied by the facor of___________ to have the sulphur content.
11. The final step in the manufacture of urea is the dehydration of __________________to form urea.
12. FAO declared 2023 as the International Year of_______________.
13. The criteria for establishing the essentiality of a nutrient element for plants were proposed by ________________________________.
14. The summative weight of all living organisms per unit mass of soil is called ____________________.
15. A group of soills of about the same age, formed from similar parent materials, and occurring under simiar climatic conditions, but having

different characteristics because of variations in relief and drainage is called a _______________.

16. A simple laboratory method for the determination of gypsum requirement of alkali soils was developed by ______________________________.
17. In Soil Taxonomy, an epipedon that is human-altered and usually high in available phosphorus is referred to as ____________________ epipedon.
18. The particular category of Soil Taxonomy represented by the term Vertic Ustorthents is _________________________.
19. The light distillate fraction of petroleum upto a boiling point of 215^0 C is called __________________.
20. The fertilizer for the manufacture of which carbon dioxide under high pressure is used is ____________________.
21. For reference in Soil Taxonomy, for the areas of Mediterranian climates where winters are moist and cool, and summers are warm and dry, the assigned Soil Moisture Regime is referred as _________________.
22. A vertical section exposing a soil is its ___________________.
23. The process in which simple inorganic forms of plant available nutrients is converted to complex organic forms unavailable to plants is called __________________________.
24. The concentration of exchangeable H^+, K^+, Na^+, Ca^{2+} and Mg^{2+} ions of a soil are 3, 39, 46, 150 and 18 mg/100 g soil, respectively. Assuming the soil not to have any other exchangeable cation, the Exchangeable Sodium Per centage (ESP) of the soil is _______________.
25. Autotrophic microorganisms derive their carbon requirement from _________________.
26. Shoemaker described a simple laboratory method for the determination of __________________________ soils.
27. The fertilizer of which nearly 70% of the gobal requirement is now met from caprolactam industry is ____________________.

28. In the layer lattice structure of montmorillonite, one aluminium octahedral sheet remains attached to two ______________ tetrahedral sheets on either side of the former.

29. The year 2015 was declared and celebrated by FAO as the International Year of ______________.

30. The mineral langbeinite is a double sulphate of potassium and ____________________ and the two respective sulphates are present in the mineral in ______ molecular ratio.

31. The process in which complex organic and plant unavailable forms of nutrients is converted to simple inorganic forms of the nutrients available to plants is called __________________________.

32. Minerologically, micas are double silicates of ____________________ and ____________________.

33. The concentration of exchangeable H^+, K^+, Na^+, Ca^{2+} and Mg^{2+} ions of a soil are 2, 58.5, 23, 160 and 18 mg/100 g soil, respectively. Assuming the soil not to have any other exchangeable cation, the Percent Base Saturation (PBS) of the soil is ________________.

34. The process of chemical weathering in which a substance absorbs moisture so that the absorbed water becomes an integral part of the molecule, often with a large increase in volume, is called ____________________.

35. The number of soil test parameters in the Soil Health Card Scheme of the Government of India is ____________.

36. For particular use in soil mapping, a Series in Soil Taxonomy is often followed by ____________________ that describes a particular property of the area such as slope, erosion, stoniness etc.

37. Soil formation process characterized by the removal of silica from the upper soil layers and its deposition in the lower layers due to which the upper soil layer becomes poor in silica and proportionately rich in sequioxided is called __________________________.

38. If the mean annual soil temperature is $< - 10^0$ C, the Soil Temperature Class (STC) used as a part of the family name of a soil is __________________________.

39. In Soil Taxonomy, diagnostic surface horizons are referred to as ________________________.

40. A freshly sampled soil mass of 126.4 g on oven drying weighed 95.8 g. Assuming the bulk density of the soil to be 1.38 g/cm^3, the per cent moisture content of the native soil on gravimetric and volumetric basis are ______________ and ______________.

41. The practice of keeping farm animals in the fallow lands after the harvest of the last crop for days and nights so as to enrich the soil with plant nutrients through their excreta is called ______________.

42. The branch of soil science concerned with the study of soils for plant growth is called ________________________.

43. A cation that can substitute K^+ for fixation in the interlayer space of illite is ________________________.

44. In wind erosion, the movement of soil particles along ground surface by short series of bounces of not more than 30 cm high is called ________________________.

45. In the determination of CEC of a soil by NH_4^+ saturation followed by NH_3 distillation method, the titre value for a KCl leachate prepared from 5 g soil was 4.7 ml 0.1 N H_2SO_4. If it is so, the CEC of the soil is ________ m.e./100 g soil.

46. According to the norms of the Soil Health Card Scheme of the Government of India, the grid area for drawing soil samples in the rainfed areas is ________ ha.

47. The study of the formation, composition, structure and properties of rocks is called ________________________

48. In the expression for the calculation of Sodium Adsorption Ratio (SAR) of irrigation water, the concentrations of the ions viz. Ca^{2+}, Mg^{2+}, K^+ and Na^+ are expressed in terms of ________________________.

49. In Soil Taxonomy, the diagnostic subsurface horizon that is loamy textured, weakly cemented and is brittle is referred by the term ________________________.

50. The volume of 0.05 N KOH solution required for the titration for complete neutralization of 5 g H^+-saturated montmorillonite (CEC 120 m.e./100 g) is _______ ml.

51. The two ions present in the soil solution that contribute to soil acidity are H^+ and _______.

52. According to Soil Taxonomy, the Order for the soils that have a permafrost at a shallow depth from the soil surface is _______________.

53. The fraction of the incident radiation that is reflected from the soil (land) surface is known as _____________________.

54. Of the twelve Soil Orders, two viz.______________________ and ___________do not have well developed master horizons.

55. The peeling away of the outer mass of rocks in the form of distinct layers due to differential heating and cooling between their outer and inner parts is referred to as ___________________________.

56. Igneous rocks are subclassified into acid, neutral, basic and ultrabasic igneous rocks depending upon their ________________ content.

57. Upper 20 miles of the lithosphere of the earth is referred to as ___________________________.

58. The concentration of exchangeable H^+, K^+, Na^+, Ca^{2+} and Mg^{2+} concentrations of a soil are 3, 39, 46, 190 and 30 mg/100 g soil, respectively. Assuming the soil not to have any other exchangeable cation, the cation exchange capacity of the soil is __________ m.e./100 g soil.

59. A colloidal solution is made up of two phases, viz. a dispersed phase and a ___________________________________.

60. The mineral that occurs in the earth as an intimate mixture of the carbonates of calcium and magnesium is ____________________.

61. The two components of ammonia buffer frequently used in many laboratory analyses such as the determination of Ca+Mg by EDTA method are ammonia and _____________________________.

62. Application of ammonium sulphate (24% S) @ 180 kg/ha adds around __________ kg of S to a field of 0.5 ha area.

63. If an analyst used 2.85 ml of a commercially available concentrated acid for preparing 1 L 0.1 N acid solution, then the analyst must have used concentrated _________________________ acid.

64. Saline alkali and alkali soils differ with regard to their ____________.

65. In Soil Taxonomy, diagnostic subsurface horizons are referred to as ______________________________.

66. The enzyme involved in biological nitrogen fixation is ______________________________.

67. The product obtained from the excrement and dead remains of some sea birds lying covered underneath the sea and ocean coast for hundreds of years that can now be used as a fertilizer is known as __________________________________.

68. The upper six to seven inches of the top soil that is of much concern to the farmers is called ______________________________.

69. The potassic fertilizer indigenously manufactured in India is ______________________________.

70. In EDTA, the total number of binding sites available for chelation with micronutrient cations in soils is ____________.

71. The law that describes the relationship between the decrease in the intensity of the transmitted radiation with an increase in the concentration of a solution is described by ________________ law.

72. There are three mechanisms through which plant nutrients in soil solution reach the surface of root hairs viz. diffusion, ____________________ and ______________________.

73. Single super phosphate fertilizer is an intimate mixture of ___________________ and __________________________.

74. An important soil property that does not change noticeably over a period of time running in decades is the soil's ______________________.

75. For perfect adherence in obeying Stokes' law, the shape of the falling particles in the viscous medium has to be _________________.

76. The phytotoxic substance often present in fertilizer grade urea the concentration of which must not exceed a prescribed limit is _________________________.

77. Softening and opening of the seed coats of seeds like those of dhaincha by mechanical means such as by light pounding with sands or thermally

or chemically for facilitation of easy germination is called ____________________________.

78. The most important source of hydrogen for the manufacture of nitrogenous fertilizers in India is ____________________________.

79. A complex fertilizer of which dicalcium phosphate is a constituent is ____________________________.

80. Atomic forms of an element that have the same atomic number but have different atomic weights are called ________________________ of each other.

81. The capacity of a moist soil to be molded into definite shapes is called the soil's ______________________.

82. Of the several Soil Temperature Regimes (STR) currently in vogue in Soil Taxonomy, the hottest one (STR) for reference in areas where the Mean Annual Summer Temperature (MAST) is 22^0 C or more, and the temperature difference between mean summer and mean winter temperature is 6^0 C or more, is referred as ____________________.

83. In symbiotic biological nitrogen fixation by *Rhizobia*, a red coloured pigment called ____________________________ protects the enzyme nitrogenase from being degraded by molecular oxygen.

84. A pH meter measures the difference in potential developed due to the difference in concentrations of ________ ion of the test solution in reference to the same of a reference setup.

85. A fertilizer that contains only one nutrient absorbable by plants is called a __________________________ fertilizer.

86. Super phosphoric acid is used in the manufacture of the fertilizer ____________________________________.

87. Ratio law governing the distribution of exchangeable ions in soil solution was first formulated by ______________________.

88. Given that the molecular weight of $FeSO_4$ is 151.85, the equivalent weight of it in the following reaction is ____________.

$$6\ FeSO_4 + K_2Cr_2O_7 + 7\ H_2SO_4 = 3\ Fe_2(SO_4)_3 + K_2SO_4 + Cr_2(SO_4)_3 + 7\ H_2O$$

89. Igneous rocks are formed by the cooling down of ____________________.

90. The fertilizer that ranks the second in terms of consumption in India is ______________________________.

91. The concentrations of Na^{+}, Ca^{2+} and Mg^{2+} of an irrigation water source are 1, 6 and 2 m.e./L. The Sodium Adsorption Ratio (SAR) of the water source is ____________.

92. Hematite on hydration changes to the mineral ______________________.

93. For preparing 0.5 L 500 ppm iron (Fe) solution, the amount of $FeSO_4$ to be dissolved in water is __________ g (atomic weight of Fe is 55.85).

94. A material – either natural or synthetic – that is spread over the soil surface to help prevent the loss of soil moisture through the process of evaporation is called a ____________.

95. In Soil Taxonomy, as one goes down in the hierarchical system, the category Suborder is followed the by category ________________.

96. Red colour of soils is mostly due to the presence of ______________________.

97. In Land Capability Classification maps, soils of Class I are shown in ____________ colour.

98. Potassium containing mineral sylvinite is a mixture of potassium chloride and ______________________________.

99. Climate apart, the other active factor of soil formation is _________________.

100. Relatively young soils formed by the transportation of soil materials by rivers and streams are called ____________________ soils.

101. Microorganisms which derive their energy requirement from the oxidation of simple inorganic compounds such as those of ammonium, sulphide etc. are called ________________________________.

102. The micronutrient element the availability of which increases with increase in soil alkalinity is __________________________.

103. Soil test results recommend the addition of nitrogen @120 kg/ha to be applied through ammonium sulphate (20.6% N) and neem cake (5% N) in 4:1 ratio. The amounts of ammonium sulphate and neem cake to be

applied to a 0.5 ha field are_____________ and ___________kg, respectively.

104. In Soil Taxonomy, Salic, Sombric and Glossic are different types of ________________________.

105. According to Soil Taxonomy, the Order to which the soils of volcanic origin belong to is _________________________.

106. The mobile application launched by the Government of India in collaboration with Indian Meteorological Department and Indian Council of Agricultural Research for helping farmers in weather-based farming is named as ____________________________.

107. The straight phosphatic fertilizer for the manufacture of which phosphoric acid is used is _______________________________.

108. The combustible gas of biogas plants is ___________________.

109. In PEC process of manufacture of nitrophosphates, rockphosphate is treated with a mixture of two acids viz. ________________________ and _______________________ acid.

110. Optimum C:N ratio of soils is ___________________.

111. Microorganism which require low temperature (< 10^0 C) for their best growth are called ______________________________.

112. In nitrification process, the *Nitrobacter* bacteria execute the oxidation of ________________________________ to __________________.

113. Statutory health authorities recommend the ________________ level in drinking water to be below 10 mg per L so as to prevent the prevalence of methemoglobinemia.

114. Polished and grooved surfaces produced by one soil mass sliding past another common in Vertisols rich in expanding type clays with high swelling and shrinkage property are called ______________________________

115. Considerable amounts of fertilizer potassium are temporarily fixed in the interlayer space of __________________ (a layer lattice silicate clay).

116. *Akiochi* disease in rice is caused due to poisoning by _________________ in waterlogged soils.

117. Soil formation process characterized by the removal of sesquioxides from the upper soil layers and its deposition in the lower layers due to which the upper soil layer becomes poor in sesquioxides and proportionately rich in silica is called ____________________.

118. The micronutrient element essential for the functioning of plastocyanin in the process of photosynthesis is ____________________.

119. Hornmeal is a concentrated organic manure containing ________________ as the major nutrient.

120. The 17th entrant to the list of nutrients essential for plants is ________________.

121. A soil with pH of 9.0, EC 4.8 dS/m at 25^0 C and ESP 22 is classified as a/an ________________ soil.

122. Soil moisture tension at field capacity in terms of pF is ________.

123. A relatively low grade water insoluble phosphatic fertilizer produced in steel plants as a byproduct is ____________________.

124. An alkali soil with a cation exchange capacity of 24 m.e./100 g soil must have an exchangeable sodium per cent (ESP) not less than ____________ mg/100 g soil.

125. Individual crystals of kaolinite clay are joined together by ______________ bonding making it a nonexpanding clay mineral.

126. In a typical neutron moisture probe, the fast moving neutrons are produced by the collision of the α particles emitted by radioactive ___________ with the relatively light nuclei of the element ______________.

127. The force of attraction between two similar particles is called _______________, while ____________________ refers to the same between two dissimilar particles.

128. The principal cation responsible for flocculation of inorganic soil silicate clays leading to a soil structure favourable for plant growth is ___________.

129. Enrichment of the water of lakes and rivers with nutrients, especially nitrogen and phosphorus, resulting in a boom in the population of algae and other aquatic species along with a concommitant sharp fall in oxygen level of the waterbody, is called ____________________.

130. The practice of dumping urban and industrial wastes of the cities and towns in low lying swampy areas with the intention of making the sites as normal upland areas in course of time is called ________________.

131. The salt chosen as the reference salt for assessing the Salt Index of fertilizers is ______________________________.

132. The unit for expressing the redox potential of chemical systems is ________.

133. Manufacture of muriate of potash from sylvinite ore invoves the separation of potassium chloride crystals from ________________ present in the ore.

134. The energy of an electromagnetic radiation is inversely related to its ______________.

135. A volume of 100 ml 10 ppm Cu solution contains ______ g $CuSO_4.5 H_2O$ (atomic weight of Cu is 63.546).

136. The most abundant of all the elements in the earth's crust is ______________.

137. Of the different constituents of organic matter such as polysachharides etc., the one that is the most resistant to microbial degradation is __________________.

138. The volume of a cylindrical core sampler with internal diameter of 10 cm and height 14 cm is ______ cm^3.

139. The source of radiation in the analytical instrument atomic absorption spectrophotometer is a ______________________________________.

140. The charcoal like substance that has much potential for use in farming and is produced by burning organic material from agricultural and forestry wastes in a controlled pyrolysis process is called __________________.

141. A common extractant for the determination of available potassium content of soils is _________________________.

142. The dispersion agent used in the mechanical analysis of soils is ______________________________.

143. In cation exchange process operating in soils, 30 mg Mg^{2+} is exchanged by _______ g NH_4^+.

144. The plant nutrient element that is intimately related to ATP in the energy transfer process in both plant and animal systems is ____________________.

145. Dry and deodourised concentrated organic manure obtained from night soil by its controlled digestion is called ______________________.

146. Nangal unit of National Fertilisers Limited used in the past ______________ as the source of hydrogen for the manufacture of ammonia and nitric acid.

147. In the determination of magnesium by EDTA method, 3.2 ml of 0.02 N EDTA equals to ________ mg magnesium.

148. The ionic form of phosphorus produced by the dissolution of dicalcium phosphate is ____________.

149. The principal soil amendment for the reclamation of alkali soils is ____________________.

150. The reagent used for the oxidation of organic carbon in the determination of organic carbon content of soils by Walkley and Black's method is ________________________________.

151. The solum of a soil together with the parent material of the soil is called ____________________.

152. The principal hydrocarbon constituent of natural gas used as a source of hydrogen for ammonia manufacture in fertilizer plants is __________________________.

153. Fertilizer industry began its run in India with the manufacture of ________________________________ in preindependent India's first fertilizer plant established at Ranipet inTamil Nadu.

154. A __________________ is a category, though not true or technical, in Soil Taxonommy on the basis of some important and purely local deviation that influences the use of the soil, such as surface texture, erosion, stoniness, or soluble salt content warranting its mention in the Soil Taxonomic name already framed.

155. Denitrifying bacteria use ________________________ as the terminal acceptor of electrons in place of molecular oxygen.

156. The letter 'g' used in describing the subordinate differentiation of a master horizon, say for instance B_g, denotes __________________.

157. Single super phosphate is manufactured by acidulation of rock phosphate with __________________ acid.

158. In Remote Sensing, the Sensor used in microwave region is called a __________________.

159. For establishing a Series in Soil Taxonomy, the minimum required area is ____________ acres.

160. In Soil Taxonomy, the Soil Moisture Regime that is intermediate between the Aridic and Udic Soil Moisture Regimes is the __________________ Soil Moisture Regime.

III. Formulae/Compositions of Minerals/Fertilizers

Furnish the formulae OR chemical compositions of the following minerals/fertilizers

1. Actinolite
2. Albite
3. Ammonium carbamate
4. Anatase
5. Andalusite
6. Anhydrite
7. Anorthite
8. Apatite
9. Azurite
10. Barite
11. Bauxite
12. Beryl
13. Biotite
14. Biuret
15. Borax
16. Bornite
17. Brucite
18. Calamine
19. Calcite
20. Carnallite
21. Chalcopyrite (coppr pyrite)
22. Chalcocite
23. Chile salt peter
24. Colemanite
25. Corundum
26. Covellite
27. Cryolite
28. Dicalcium phosphate
29. Dolomite
30. Emerald
31. Enstatite
32. Epidote
33. Epsomite
34. Fayalite
35. Ferrosilite
36. Fluorapatite

37. Fluorspar
38. Forsterite
39. Francolite
40. Friedel salt
41. Galena
42. Garnet
43. Gibbsite
44. Glauber's salt
45. Goethite
46. Gypsum
47. Halite
48. Hausmannite
49. Hematite
50. Ilmenite
51. Inderite
52. Kainite
53. Kaolinite
54. Kieserite
55. Kyanite
56. Langbeinite
57. Leucite
58. Limonite
59. Magnesite
60. Magnetite
61. Malachite
62. Manganite
63. Mohrite
64. Molybdenite
65. Muscovite
66. Nepheline
67. Nitre
68. Orthoclase
69. Oxyapatite
70. Perovskite
71. Polyhalite
72. Potassium schoenite
73. Powellite
74. Pyrite

75. Pyrolusite
76. Quartz
77. Rhodochrosite
78. Rutile
79. Sanidine
80. Serpentine
81. Siderite
82. Single super phosphate
83. Smithsonite
84. Sphalerite
85. Spinel
86. Stengite
87. Super phosphoric acid
88. Sylvinite
89. Sylvite
90. Talc
91. Thiourea
92. Titanite/Sphene
93. Topaz
94. Triple super phosphate
95. Variscite
96. Vivianite
97. Wollastonite
98. Wulfenite
99. Zinc blende
100. Zircon

IV. Contributions of Scientists in Soil Science

Mention the major contribution/s of the following scientists in Soil Research/Soil Science

1. J. Tyndall
2. C. F. Marbut
3. G. W. Robinson
4. A. H. Munsell
5. E. W. Hilgard
6. D. I. Arnon and P. R. Stout
7. Dr C. N. Acharya
8. Guy D. Smith
9. A. Howard and Y. D. Wad
10. Hans Jenny
11. R. K. Schofield
12. Thomas Graham
13. H. P. J. Darcy
14. Friedrich Wohler
15. Friedrich Mohs
16. H. E. Shoemaker
17. J. H. Gilbert
18. J. B. Lawes
19. J. Samuel Joffe
20. G. G. Stokes
21. V. V. Dokuchaev
22. Robert Brown
23. J. W. Leather
24. T. C. Chamberlin and F. R. Moulton
25. H. B. Hutchinson and E. H. Richards
26. H. Lundegardh
27. E. R. Beaufils
28. S. P. L. Sorenson
29. Justus von Leibig
30. Thomas Way
31. Jean Baptiste Boussingault
32. Adolf Fick
33. M. W. Beijerinck
34. W. R. Schoonover
35. A. M. Atterberg
36. L. A. Richards
37. S. Winogradsky
38. Theodore de Schwann
39. Fritz Haber and Carl Bosch
40. Fowler and Ridge
41. C. C. Nikiforoff
42. Konstantin Glinka
43. R. B. Cate and L. A. Nelson
44. George N. Coffey
45. Friedrich Albert Fallou
46. Milton Whitney
47. C. G. Hopkins
48. R. Warrington

V. Definitions - Explanations - Lucid Notes

Define or explain or write a lucid note on the followings

1. Acid
2. Acid Equivalentof fertilizers
3. Acid rain
4. Acid sulphate soils
5. Activated Sludge Process
6. Active acidity of soils
7. Adhesion
8. Adsorption
9. Albedo
10. Alkalization of soils
11. Alluvial soils
12. Ammonification
13. Anion Exchange Capacity of soils
14. Atomic Absorption Spectrophotometry
15. Atterberg's Limits
16. Base
17. Base Equivalent of fertilizers
18. Bedrock
19. *Bhangar* and *Khadar*
20. Biological Nitrogen Fixation
21. Biological Oxygen Demand
22. Biological weathering
23. Bioremediation
24. Black soils
25. Brownian movement
26. Buffer
27. Bulk Density of soils
28. Carbon Farming
29. Carbonation
30. Catena
31. Cation Exchange Capacity of soils
32. C-Axis spacing
33. Chelate
34. Chemical Oxygen Demand
35. Chemical weathering
36. Chemoautotrophs
37. Chernozem soils
38. Chesnutt soils
39. Chroma of colours
40. Chromatography
41. Coefficient of Linear Extensibility of soils
42. Cohesion
43. Colloid
44. Colluvium
45. Colorimetry
46. Complementary ions
47. Critical Level of nutrients in plants
48. Critical Relative Humidity
49. Darcy's law
50. Deflocculation
51. Denitrification
52. Desert/Arid soils
53. Desertification

54. Diagenesis
55. Diagnostic horizon
56. Diffuse Double Layer
57. Diffusion
58. DRIS
59. *Dubba* and *Chalka* soils
60. Dunes
61. Earth's Crust
62. Edaphology
63. Electromagnetic Spectrum
64. Eluviation
65. Endopedon
66. Epipedon
67. Equilibrium Constant and Law of Mass Action
68. Equivalent weight
69. Equivalent weight of acids
70. Equivalent weight of bases
71. Essential minerals
72. Eutrophication
73. Exchangeable Sodium Per cent
74. Exfoliation
75. Extrusive rocks
76. Fick's law of Diffusion
77. Field capacity
78. Flame photometry
79. Flocculation
80. Fourier's law
81. Gapon equation
82. Gleization
83. Gypsum Requirement of alkali soils
84. Hardy Schulze rule
85. Heavy metals
86. Hidden hunger and Luxury consumption of plant nutrients
87. Hue of colours
88. Humus
89. Hydration
90. Hydraulic conductivity
91. Hydrolysis
92. Hygroscopic coefficient
93. Igneous rocks
94. Illuviation
95. Immobilization
96. Intrusive rocks
97. Ion Exchange Capacity
98. Ionic Product of water
99. Jarusov's rule
100. Lambert Beer's law
101. Land Capability Classification
102. Land Degradation
103. Landfills
104. Laterite and lateritic soils
105. Laterization
106. Leaching Requirement of saline soils
107. Lime Potential
108. Lime Requirement of acid soils
109. Lithosphere

110. Loess
111. Lyotropic series
112. Magma
113. Mass flow
114. Melanization
115. Metamorphic rocks
116. Mineral
117. Mineralization
118. Mohs scale
119. Moist bolus
120. Mole
121. Mole fraction
122. Mor
123. Mulching
124. Mull
125. Mycorrhizae
126. Nano fertilizer
127. Nitrification
128. Origin of negative charge in humus
129. Origin of negative charge in silicate clays
130. Origin of positive charge in silicate clays
131. Oxidation
132. Oxidizing agent
133. Parent material of soils
134. Particle Density of soils
135. Pedalfer
136. Pedocal
137. Pedology
138. Pedon
139. Pedoturbation
140. Penning
141. Per cent Base Saturation
142. Permafrost
143. Permeability Index of irrigation water
144. Petrology
145. pF
146. pH
147. Phosphate Potential
148. Phyllosilicates
149. Physical weathering
150. Plasticity
151. Plutonic rocks
152. Podzolization
153. Poiseuille's law
154. Potassium Adsorption Ratio
155. Potential Buffering Capacity of soils
156. Poudrette
157. Primary minerals
158. Primary standard substance
159. Q/I Relationship
160. Radioactivity
161. Radioactivity – Units
162. Rainwater harvest
163. Ratio law
164. Red soils
165. Redox Potential
166. Reducing agent

167. Reduction
168. Regolith
169. Regosol
170. Regur
171. Relief as a soil forming factor
172. Reserve acidity of soils
173. Residual Sodium Bicarbonate of irrigation water
174. Residual Sodium Carbonate of irrigation water
175. Rocks
176. Rubification
177. Runoff
178. Salinization of soils
179. Salt Affected Soils
180. Salt Index of fertilizers
181. Salt Index of irrigation water
182. Saltation (in wind erosion)
183. Scarification
184. Secondary minerals
185. Sedimentary rocks
186. Sewage
187. Sewerage
188. Shelterbelt
189. Slickensides
190. Sludge
191. Sodium Adsorption Ratio
192. Soil Amendments
193. Soil Conditioners
194. Soil Consistence
195. Soil Crusting
196. Soil Horizon
197. Soil Permeability
198. Soil Profile
199. Soil Structure
200. Soil Taxonomy
201. Soil Taxonomy – Categories
202. Soil Taxonomy – Nomenclature
203. Soil Taxonomy – Soil Moisture Regimes
204. Soil Taxonomy – Soil Temperature Regimes
205. Soil Texture
206. Soils of India
207. Solonchak soils
208. Solonetz soils
209. Solubility Product
210. Solum
211. Spectrophotometry
212. Standard solution
213. Sticky point
214. Stokes' law
215. Sullage
216. Surface creep (in wind erosion)
217. Suspension (in wind erosion)
218. Suspension effect (in soil pH measurement)
219. Sward
220. Topography
221. Tyndall effect
222. Universal Soil Loss Equation

223. Value of colours
224. Vermicomposting
225. Vermiwash
226. Volcanic rocks
227. Watershed
228. Weatherability of Minerals
229. Weathering
230. Wetlands
231. Wilting point
232. Wind Erosion Equation
233. Windbreak
234. Zeta Potential

VI. Expressions for Deriving Numerical Values of Soil Characteristics

Furnish the formulae for the numerical calculation of the following soil properties/characteristics

1. Bulk density
2. Particle density
3. Porosity
4. Per cent Pore space
5. Per cent Base Saturation (PBS)
6. Exchangeable Sodium Percentage (ESP)
7. Sodium Adsorption Ratio (SAR)
8. Soluble Sodium Per cent of irrigation water
9. Potassium Adsorption Ratio (KAR)
10. Residual Sodium Carbonate (RSC)
11. Residual Sodium Bicarbonate (RSBC)
12. Coefficient of Linear Extensibility (COLE)
13. Stokes' law
14. Darcy's law
15. Fick's Law of Diffusion
16. Leaching Requirement for reclamation of saline soils
17. Permeability Index of irrigation water

VI. Expressions for Deriving Numerical Values of Soil Characteristics

VIII. Answers

I. Choosing the Right Answer Option

Underline the most appropriate answer (* solutions for the Questions with * mark is worked out in brief at the end of the answers)

1. (b)	2. (d)	3. (d)	4. (c)	5. (d)
6. (a)	7. (b)	8*. (d)	9. (c)	10. (c)
11. (a)	12. (d)	13. (b)	14. (d)	15. (c)
16. (d)	17. (b)	18. (b)	19. (c)	20. (d)
21. (c)	22. (a)	23. (b)	24. (b)	25. (a)
26. (c)	27. (c)	28. (b)	29. (c)	30. (a)
31. (b)	32. (c)	33. (b)	34. (b)	35. (c)
36. (b)	37. (c)	38. (a)	39. (a)	40. (a)
41. (d)	42. (a)	43. (c)	44. (a)	45. (b)
46. (d)	47. (d)	48. (c)	49. (b)	50*. (a)
51. (a)	52. (a)	53. (a)	54. (c)	55. (a)
56. (b)	57. (c)	58. (d)	59. (c)	60. (a)
61. (b)	62*. (c)	63. (a)	64. (d)	65*. (c)
66. (b)	67. (c)	68. (c)	69. (b)	70. (d)
71. (b)	72. (b)	73. (a)	74. (c)	75. (a)
76. (d)	77. (c)	78. (a)	79. (b)	80. (b)
81. (c)	82. (c)	83. (d)	84. (d)	85. (c)
86. (b)	87. (d)	88. (d)	89. (d)	90. (a)

91. (c)	92. (b)	93. (c)	94*. (b)	95*. (c)
96. (b)	97. (c)	98. (d)	99. (b)	100*. (d)
101*. (c)	102. (c)	103. (a)	104. (d)	105. (b)
106. (b)	107. (b)	108. (a)	109. (d)	110. (d)
111. (d)	112. (b)	113. (b)	114. (a)	115. (b)
116. (a)	117. (b)	118. (c)	119. (c)	120. (d)
121. (b)	122. (b)	123. (d)	124. (b)	125. (b)
126. (c)	127. (d)	128. (c)	129. (c)	130. (b)
131. (c)	132. (b)	133. (d)	134. (c)	135. (a)
136. (a)	137. (b)	138. (b)	139. (c)	140. (c)
141. (d)	142. (a)	143. (b)	144. (b)	145. (b)
146. (d)	147. (c)	148. (d)	149. (b)	150. (d)
151. (a)	152. (b)	153. (d)	154. (b)	155. (d)
156. (a)	157. (c)	158. (b)	159. (d)	160. (b)
161. (d)	162. (a)	163. (b)	164. (b)	165. (b)
166. (c)	167. (c)	168*. (c)	169. (b)	170. (a)
171. (d)	172. (a)	173. (d)	174. (c)	175. (a)
176. (b)	177. (c)	178. (d)	179. (d)	180. (d)
181. (d)	182. (b)	183. (c)	184. (b)	185. (d)
186. (d)	187. (a)	188. (c)	189. (c)	190. (a)

191. (a)	192. (b)	193. (a)	194. (c)	195. (c)
196. (b)	197. (b)	198. (c)	199*. (d)	200. (a)
201. (b)	202. (b)	203. (a)	204. (a)	205. (d)
206. (a)	207. (a)	208. (a)	209. (c)	210*. (b)
211. (d)	212. (a)	213. (b)	214. (c)	215. (a)
216. (c)	217. (c)	218*. (c)	219. (b)	220. (d)
221. (a)	222. (d)	223. (c)	224. (d)	225. (d)
226. (c)	227. (c)	228. (b)	229. (a)	230. (b)
231. (d)	232. (d)	233. (d)	234. (c)	235*. (c)
236. (c)	237. (a)	238. (a)	239. (d)	240. (a)
241. (d)	242. (d)	243. (d)	244. (b)	245. (c)
246. (d)	247. (b)	248. (d)	249. (a)	250. (a)
251. (c)	252. (b)	253. (a)	254*. (b)	255*. (c)
256. (c)	257. (a)	258. (d)	259. (d)	260. (c)
261. (b)	262. (c)	263. (b)	264. (d)	265. (b)
266. (c)	267. (c)	268. (b)	269. (c)	270. (d)
271. (d)	272. (a)	273. (b)	274. (d)	275. (c)
276. (b)	277. (a)	278. (d)	279. (b)	280. (b)
281. (c)	282. (a)	283. (c)	284. (c)	285. (d)
286. (a)	287. (d)	288. (c)	289. (b)	290. (d)

291. (b)	292. (b)	293. (a)	294. (b)	295. (d)
296. (d)	297. (a)	298. (d)	299. (d)	300. (b)
301. (a)	302. (c)	303. (a)	304. (d)	305. (a)
306. (a)	307. (d)	308. (c)	309. (c)	310. (c)
311. (b)	312. (a)	313. (c)	314. (a)	315. (b)
316. (d)	317. (a)	318. (d)	319. (c)	320. (c)
321. (a)	322. (c)	323. (d)	324. (a)	325. (a)
326*. (c)	327. (d)	328. (b)	329. (b)	330. (c)
331. (d)	332. (c)	333. (b)	334. (a)	335. (b)
336. (d)	337. (b)	338. (b)	339. (c)	340. (d)
341. (b)	342. (c)	343. (a)	344. (a)	345. (a)
346. (c)	347. (d)	348. (d)	349. (a)	350. (b)
351. (b)	352. (b)	353. (c)	354. (c)	355. (b)
356. (d)	357. (c)	358. (b)	359. (c)	360. (d)
361. (a)	362. (a)	363. (a)	364. (b)	365. (a)
366. (d)	367. (d)	368. (d)	369. (c)	370. (c)
371. (a)	372*. (c)	373. (c)	374. (d)	375. (b)
376. (d)	377. (d)	378. (d)	379. (b)	380. (c)
381. (d)	382. (b)	383. (c)	384. (a)	385. (b)
386. (a)	387. (d)	388. (d)	389. (d)	390. (c)

391. (a)	392. (c)	393. (a)	394. (c)	395. (b)
396. (d)	397. (c)	398. (b)	399. (a)	400. (c)
401. (a)	402. (b)	403. (c)	404. (a)	405. (c)
406. (a)	407. (d)	408. (c)	409. (d)	410. (c)
411.(d)	412. (b)	413. (d)	414. (c)	415. (d)
416. (c)	417. (c)	418. (a)	419. (d)	420. (a)
421. (b)	422. (a)	423. (d)	424*. (c)	425. (b)
426. (a)	427. (b)	428. (d)	429. (a)	430. (b)
431. (c)	432. (b)	433. (d)	434. (b)	435. (a)
436. (d)	437. (d)	438. (a)	439. (d)	440. (c)
441. (d)	442. (a)	443. (a)	444. (b)	445. (d)
446. (c)	447. (d)	448. (b)	449. (a)	450. (d)
451. (b)	452. (c)	453. (d)	454. (d)	455. (a)
456. (c)	457. (a)	458. (a)	459. (a)	460. (d)
461. (c)	462. (a)	463. (b)	464. (d)	465. (d)
466. (a)	467. (a)	468. (a)	469. (b)	470. (d)
471. (d)	472. (c)	473. (d)	474. (d)	475. (d)
476. (b)	477*. (c)	478. (c)	479. (b)	480. (a)
481. (a)	482. (a)	483. (d)	484. (b)	485. (a)
486. (b)	487. (c)	488. (d)	489. (a)	490. (b)

491. (a)	492. (a)	493. (d)	494. (d)	495. (d)
496. (b)	497. (a)	498. (c)	499. (b)	500. (b)
501. (d)	502. (d)	503. (d)	504. (c)	505. (b)
506. (d)	507. (d)	508. (d)	509. (d)	510. (d)
511. (d)	512. (d)	513. (b)	514. (d)	515. (a)
516. (d)	517. (c)	518. (c)	519. (b)	520. (c)
521*. (d)	522. (a)	523. (b)	524. (a)	525. (b)
526. (c)	527. (a)	528. (d)	529. (d)	530. (a)
531. (d)	532. (d)	533. (a)	534. (a)	535. (d)
536. (a)	537. (d)	538. (b)	539. (d)	540. (c)
541. (c)	542. (d)	543. (b)	544. (d)	545. (b)
546. (b)	547. (b)	548. (c)	549. (c)	550. (a)
551. (c)	552. (d)	553. (b)	554. (d)	555. (c)
556. (d)	557. (d)	558. (a)	559. (d)	560. (d)
561. (b)	562. (c)	563. (a)	564. (d)	565. (b)
566. (a)	567. (b)	568. (c)	569. (a)	570. (c)
571. (a)	572. (b)	573. (d)	574. (a)	575. (a)
576. (d)	577. (c)	578. (b)	579. (c)	580. (d)
581. (a)	582. (c)	583. (a)	584. (a)	585. (b)
586. (d)	587. (c)	588. (c)	589*. (c)	590. (d)

591. (c)	592. (b)	593. (c)	594. (b)	595. (c)
596. (c)	597. (b)	598. (b)	599. (a)	600. (a)
601. (c)	602. (d)	603.(c)	604. (d)	605. (d)
606. (a)	607. (b)	608. (c)	609. (b)	610. (a)
611. (a)	612. (b)	613. (b)	614. (a)	615. (c)
616. (d)	617. (b)	618. (d)	619. (d)	620. (d)
621. (a)	622. (b)	623. (c)	624. (d)	625. (a)
626. (a)	627. (b)	628. (c)	629. (c)	630. (b)
631. (a)	632. (c)	633. (b)	634. (d)	635. (b)
636. (b)	637. (d)	638. (d)	639. (d)	640. (b)
641. (c)	642. (b)	643. (c)	644. (d)	645. (b)
646. (d)	647. (c)	648. (b)	649. (d)	650. (b)
651. (d)	652. (d)	653*. (c)	654. (b)	655. (d)
656. (c)	657. (a)	658. (b)	659. (a)	660. (b)
661. (d)	662. (a)	663. (d)	664. (a)	665. (c)
666. (b)	667. (d)	668. (a)	669. (b)	670. (c)
671. (d)	672. (b)	673. (b)	674. (c)	675. (a)
676. (c)	677. (d)	678. (d)	679. (b)	680. (c)
681. (b)	682. (a)	683. (d)	684. (b)	685. (a)
686. (a)	687. (d)	688. (a)	689. (c)	690. (b)

691. (d)	692. (b)	693. (d)	694*. (b)	695. (a)
696. (b)	697. (c)	698. (d)	699*. (d)	700. (c)
701. (b)	702. (a)	703. (a)	704. (b)	705. (a)
706. (d)	707. (b)	708. (a)	709. (c)	710. (a)
711. (a)	712. (a)	713. (c)	714. (b)	715. (c)
716. (d)	717. (a)	718. (b)	719. (b)	720. (d)
721. (b)	722. (d)	723. (c)	724. (b)	725. (c)
726. (a)	727. (d)	728. (c)	729. (b)	730. (a)
731. (c)	732. (a)	733. (d)	734. (b)	735. (a)
736. (c)	737. (d)	738. (d)	739*. (c)	740. (b)
741. (d)	742. (d)	743. (a)	744. (d)	745. (c)
746. (b)	747. (b)	748. (c)	749. (d)	750. (a)
751. (b)	752. (a)	753. (d)	754. (c)	755. (a)
756. (d)	757. (b)	758*. (b)	759. (a)	760. (c)
761. (d)	762. (a)	763. (c)	764. (a)	765. (a)
766. (c)	767. (d)	768. (a)	769. (c)	770. (d)
771. (d)	772. (a)	773. (a)	774. (d)	775. (b)
776. (d)	777. (b)	778. (c)	779. (c)	780. (c)
781. (b)	782. (c)	783. (b)	784. (d)	785*. (c)
786. (d)	787. (a)	788. (b)	789. (b)	790. (b)

791. (b)	792. (a)	793. (d)	794. (c)	795. (d)
796. (d)	797. (c)	798*. (b)	799. (d)	800. (d)
801. (c)	802. (b)	803. (a)	804. (c)	805. (c)
806. (c)	807. (a)	808. (d)	809. (a)	810. (b)
811. (d)	812. (b)	813. (a)	814. (d)	815. (c)
816. (d)	817. (c)	818. (c)	819. (d)	820*. (b)
821. (b)	822. (d)	823. (a)	824. (b)	825. (d)
826. (b)	827. (d)	828. (a)	829. (c)	830*. (b)
831. (c)	832. (b)	833. (b)	834. (c)	835. (d)
836. (d)	837. (d)	838. (d)	839. (b)	840*. (c)
841. (c)	842. (c)	843. (c)	844*. (b)	845. (d)
846. (b)	847. (b)	848*. (d)	849.(a)	850. (a)
851. (d)	852*. (d)	853. (a)	854. (d)	855. (d)
856*. (d)	857. (d)	858. (a)	859. (a)	860. (d)
861. (c)	862. (d)	863. (b)	864. (b)	865. (b)
866. (d)	867. (d)	868. (c)	869. (c)	870. (c)
871. (c)	872. (d)	873. (d)	874. (a)	875*. (c)
876*. (c)	877. (b)	878. (d)	879. (a)	880. (a)
881. (b)	882. (d)	883. (b)	884. (d)	885. (a)
886. (c)	887. (a)	888. (a)	889. (d)	890. (b)

891. (a)	892. (a)	893. (c)	894. (d)	895. (c)
896. (a)	897. (b)	898. (c)	899. (b)	900. (a)
901. (a)	902. (d)	903*. (b)	904. (b)	905. (b)
906. (d)	907*. (a)	908. (c)	909. (c)	910. (b)
911. (a)	912. (d)	913. (a)	914. (d)	915. (b)
916. (b)	917. (c)	918. (d)	919. (c)	920. (d)
921. (a)	922. (c)	923. (c)	924. (d)	925. (c)
926. (d)	927. (c)	928. (d)	929. (a)	930*. (d)
931. (d)	932*. (b)	933. (c)	934. (b)	935. (a)
936. (c)	937. (a)	938. (c)	939*. (c)	940. (a)
941. (b)	942. (c)	943. (a)	944. (d)	945. (b)
946. (d)	947. (b)	948. (b)	949. (c)	950. (a)
951*. (c)	952. (c)	953. (b)	954*. (b)	955. (a)
956. (c)	957. (d)	958. (b)	959. (c)	960. (d)
961. (c)	962. (b)	963. (a)	964. (d)	965. (d)
966. (c)	967. (a)	968. (c)	969. (d)	970. (a)
971. (a)	972. (b)	973. (b)	974*. (b)	975. (a)
976. (a)	977. (c)	978. (b)	979. (b)	980. (d)
981. (b)	982. (c)	983. (d)	984. (b)	985. (b)
986. (c)	987. (a)	988. (b)	989. (a)	990. (b)

991. (b)	992. (a)	993. (a)	994. (d)	995. (a)
996. (b)	997. (d)	998. (c)	999. (c)	1000. (c)
1001. (b)	1002. (c)	1003. (d)	1004. (c)	1005. (b)
1006. (b)	1007. (a)	1008. (d)	1009. (d)	1010. (b)
1011*. (d)	1012. (d)	1013. (b)	1014. (a)	1015. (d)
1016. (b)	1017. (d)	1018. (c)	1019. (a)	1020. (b)
1021. (b)	1022. (d)	1023. (c)	1024. (a)	1025. (c)
1026*. (d)	1027. (c)	1028. (b)	1029. (a)	1030*. (b)
1031. (b)	1032. (d)	1033. (c)	1034. (a)	1035. (b)
1036. (c)	1037. (a)	1038. (b)	1039. (d)	1040. (c)
1041. (d)	1042. (a)	1043. (c)	1044. (d)	1045. (d)
1046*. (b)	1047. (d)	1048. (d)	1049. (c)	1050. (c)

Solutions for the Questions with * mark

8*. One ha = 10000 m^2
∴ 0.5 ha = 5000 m^2 = 5000 x 10000 cm^2
∴ Volume of water for 0.5 ha land with 5 cm depth is 5000 x 10000 x 5 cm^3
= $(25 \times 10^7/10^3)$ L = 250000 L

50*. Porosity, B.D. and P.D. are interrelated as,
% Pore space = $100\ (1 - \frac{B.D.}{P.D.})$
Since, B.D. and P.D. are 1.4 and 2.8 g/cm^3,
% Pore space = $100\ (1 - \frac{1.4}{2.8}) = 50$
Now, porosity is increased to 75%.
∴ The altered B.D. is, $75 = 100\ (1 - \frac{B.D.}{2.8}) = 0.7 g/cm^3$

∴ B.D. is reduced from 1.4 to 0.7 g/cm^3

∴ Upon increasing the % pore space from 50 to 75%, the B.D. of the soil is reduced to half of its original B.D.

62*. Note that, 4 milli equivalents of $H^+ \equiv$ 4 milli equivalents of $CaCO_3$

Equivalent weight of $CaCO_3$ is = molecular weight/2 = 100/2 = 50

∴ 4 milli equivalents of $H^+ \equiv$ 4 milli equivalents of $CaCO_3$ = 4 x 50 mg of $CaCO_3$ = 200 mg of $CaCO_3$

65*. Exchangeable Na content is 34.5 mg/100 g soil and CEC of the soil is 8 cmol (p^+)/kg soil (or, 8 m. e./100 g soil).

Equivalent weight of Na is 23. So, exchangeable Na content is 34.5/23 = 1.5 m.e./100 g soil

∴ ESP = (1.5/8) x100 = 18.75

∴ The soil does have pH > 8.5, EC < 4 dS/m and ESP > 15. Therefore, the soil is an alkali or sodic soil.

94*. One ha requires 250 kg urea (46% N). So, 0.5 ha requires 125 kg urea.

125 kg urea = 125 x (46/100) = 57.5 kg N

57.5 kg N = (100/25) x 57.5 = 230 kg CAN (25% N)

∴ In the present situation, due to nonavailability of urea, the farmer has to apply 230 kg CAN to his 0.5 ha field.

95*. Atomic weights of P and O are 31 and 16, respectively.

∴ 62 parts P = 142 parts of P_2O_5

∴ 1 part of P = 142/62 = 2.29 parts of P_2O_5

∴ % P_2O_5 content = % P content x 2.29

100*. One micron = 1 micrometer = 1 μm = 10^{-6} m

One milli micron = 10^{-3} micron = 10^{-3} x 10^{-6} m = 10^{-9} m

---

101*. Rate of application of N is 200 kg/ha. Field size is 0.5 ha.

∴ 0.5 L ha land requires 100 kg N.

100 kg N is to be applied in 1:3 ratio from neem cake (5% N) and urea (46% N).

∴ 0.5 L ha land requires (1/4) x 100 = 25 kg N from neem cake and (3/4) x 100 = 75 kg N from urea.

25 kg N = (100/5) x 25 = 500 kg neem cake

75 kg N = (100/46) x 75 = 163 kg urea

∴ 0.5 L ha field requires 163 kg urea and 500 kg neem cake for applying N @ 200 kg/ha with the condition of adding N from neem cake and urea in 1:3 ratio.

168*. Exchangeable Ca content is 120 mg/100 g soil. CEC of the soil is 8 cmol (p^+)/kg soil = 8 m. e./100 g soil.
Equivalent weight of Ca is 20. So, Ca content is 120/20 = 6 m.e./100 g soil
∴ Ca^{2+} saturation of the soil = [(6/8) x100] % = 75%

199*. Calculate ESP of the soil. Exchangeable sodium content is 45 mg/100 g soil = $\frac{45}{23}$ = 1.96 m.e./100 g soil.

CEC of the soil is 18 m.e./100 g soil. Therefore, ESP of the soil is $\frac{1.96 \text{ m.e.}}{18 \text{ m.e.}}$ x 100 = 11.

Since, soil pH is < 8.5, EC is > 4 dS/m and ESP is < 15, the soil in question is a saline soil.

210*. Porosity, B.D. and P.D. are interrelated as,
% Pore space = $100\ (1-\frac{\text{B.D.}}{\text{P.D.}})$
In the present case, P.D. is 2.6 g/cm^3, % Pore space is 45.
∴ B.D. is, 45 = $100\ (1-\frac{\text{B.D.}}{2.6})$
Inserting the values of P.D. and % pore space, we have, B.D. = 1.43 g/cm^3

218*. Porosity, B.D. and P.D. are interrelated as,
% Pore space = $100\ (1-\frac{\text{B.D.}}{\text{P.D.}})$
In the present case, B.D. is 1.43 g/cm^3, % Pore space is 45.
∴ P.D. is, 45 = $100\ (1-\frac{1.43}{\text{P.D.}})$ = g/cm^3
Inserting the values of B.D. and % pore space, we have, P.D. = 2.6 g/cm^3

235*. Porosity, B.D. and P.D. are interrelated as,
% Pore space = $100\ (1-\frac{\text{B.D.}}{\text{P.D.}})$
In the present case, B.D. and P.D. are 1.43 and 2.6 g/cm^3
∴% Pore space is = $100\ (1-\frac{1.43}{2.6})$
Inserting the values of B.D. and P.D., we have % pore space = 45

254*. Nitrogen content of the soil is 0.065%, while C:N ratio of the soil is 10 to 15:1 (10:1 to 15:1)

Multiplying % N content (0.065) by 10 and 15, we have % organic C content of the soil is 0.650 to 0.975% (since, C:N ratio of the soil is 10 to 15:1).

Organic C to Organic matter conversion factor is 1.724.

∴ Organic matter content of the soil is 0.650 x 1.724% = 1.121 to 0.975 x 1.724% = 1.681.

Or, organic matter content of the soil is 1.121 to 1.681%

255*. pOH of the solution is 4

∴ pH of the solution is 14 – 4 = 10

$pH = -\log[H^+]$

Or, $10 = -\log[H^+]$

Or, $[H^+]$ = antilog -10 = 10^{-10}

∴ H^+ concentration of the solution of which pOH is 4 is 10^{-10} mole per L

326*. Volume of a cylinder is given by the formula $\pi r^2 h$.

In the present case, r = 6 cm, h = 14 cm and mass of the oven dried soil is 2.2176 kg (2217.6 g)

∴ Volume of the cylindrical core sampler = $\pi r^2 h$ = (22/7) x 6^2 x 14 = 1584 cm^3

∴ Volume of the oven dried soil is 1584 cm^3

∴ B.D. of the soil is, (mass of the soil in g/volume of the in cm^3) = (2217.6 g/1584 cm^3) = 1.4 g/cm^3

372*. The carbon content of soil organic matter is around 58 per cent

∴ Organic carbon content of a soil is multiplied by the factor of (100/58) = 1.724 to have an approximate estimate of the organic matter content of a soil.

424*. Note that the reducing agent $MnSO_4$ (molecular weight, 151.0) is getting oxidised to $KMnO_4$

This implies that 1 molecule of $MnSO_4$ is getting oxidized to 1 molecule of $KMnO_4$

Oxidation states of Mn (the only element which undergoes change in O.S. in the reaction) in $MnSO_4$ and $KMnO_4$ are +2 and +7, respectively.

Change in O.S. is '+2 to +7' = 5

∴ Equivalent weight of $MnSO_4$ = Molecular weight of $MnSO_4/5$ = 151/5 = 30.2

477*. Gypsum ($CaSO_4.2\ H_2O$, molecular weight 172.2) is an acid base substance and being acidic, gypsum reacts with the base NaOH as follows

$CaSO_4.2\ H_2O + 2\ NaOH = Na_2SO_4 + 2\ H_2O + Ca(OH)_2$

The reaction states that 1 mole of gypsum reacts with 2 moles or 2 equivalents of NaOH

∴ One mole of gypsum = 2 equivalents of gypsum

∴ One mole of gypsum = 172.2 g gypsum = 2 equivalents of gypsum

∴ Equivalent weight of gypsum = 172.2/2 = 86.1

521*. Half-life ($t_{1/2}$) of P^{32} is 14.2 days (for simplicity in calculation, let us take it as 14 days). And the present radioactivity of the sample owing to P^{32} is 40000 dps. We have to find out the radioactivity of the sample in terms of dps after 70 days.

Note that, n (number of half-lives) = $t_{1/2}$ = 70/14 =5 half lives

Radioactivity in terms of dps after 5 half-lives (70 days),

$R = R^0/2^n$, where $n = 5$

∴ $R = R^0/2^5 = 40000/32 = 1250$

∴ Radioactivity of the plant sample (with initial activity of 40000 dps) after 70 days is 1250 dps

---

589*. H_2SO_4 is a dibasic acid which means that 1 mole of H_2SO_4 = 2 gram equivalents of H_2SO_4

1 mole of H_2SO_4 = 98 g H_2SO_4

∴ Equivalent weight of H_2SO_4= Molecular weight/2 = 98/2 = 49

653*. Porosity, B.D. and P.D. are interrelated as,

% Pore space = 100 $(1-\frac{B.D.}{P.D.})$

In the present case, B.D. and P.D. are 1.4 and 2.8 g/cm^3

∴ % pore space is = 100 $(1-\frac{1.4}{2.8})$ = 50

Now, B.D. is increased from 1.4 to 2.1 g/cm^3

Inserting the values of B.D. and P.D., we have

% pore space = 100 $(1-\frac{B.D.}{P.D.})$ = 100 $(1-\frac{2.1}{2.8})$ = 25

∴ Consequent to increase in B.D. from 1.4 to 2.1 g/cm^3, % pore space of the soil is reduced from 50 to 25 i. e. reduced to its half originally present.

694*. The concentration of OH^- ion of the solution is 10^{-3} mole/L.

Note that in any solution, $[H^+][OH^-] = 10^{-14}$

∴ The concentration of H^+ ion of the solution is $10^{-14}/10^{-3} = 10^{-11}$ mole/L.

∴ pH of the solution = -log 10^{-11} = 11

Alternatively, pOH = -log 10^{-3} = 3; therefore, pH = 14 - pOH = 14 - 3 = 11

699*. The pOH of the solution is 4.

Note that, pH + pOH =14

∴ pH of the solution = 14 – pOH = 14 – 4 = 10

739*. One m^2 area requires 90 g Ca for liming

Note that 1 ha = 10,000 m^2, molecular weight of limestone ($CaCO_3$) = 100, purity of limestone = 90%, and 1 ton = 1000 kg

∴ 1 ha land requires = 90 x 10,000 g Ca = 90 x 10,000 x (100/40) g pure limestone =

90 x 10,000 x (100/40) x (100/90) g 90 % pure limestone =

90 x 10,000 x (100/40) x (100/90) x 10^{-3} kg 90 % pure limestone =

90 x 10,000 x (100/40) x (100/90) x 10^{-3} x 10^{-3} ton 90 % pure limestone =

2.5 ton 90% pure limestone.

758*. One m^2 alkali soil requires 80 g Ca for reclamation

Note that 1 ha = 10,000 m^2, molecular weight of gypsum ($CaSO_4.2\ H_2O$) = 172, purity of gypsum is 95%, and 1 ton = 1000 kg

∴ 1 ha land requires = 80 x 10,000 g Ca = 80 x 10,000 x (172/40) g pure gypsum =
80 x 10,000 x (172/40) x (100/95) g 95 % pure gypsum =
80 x 10,000 x (172/40) x (100/90) x 10^{-3} kg 95 % pure gypsum =
80 x 10,000 x (172/40) x (100/95) x 10^{-3} x 10^{-3} ton 95 % pure gypsum = 3.62 ton 95% pure gypsum.

785*. Two milliequivalents of SO_4^{2-} require two milliequivalents of NO_3^- for replacement.
Equivalent weight of NO_3^- is 62
∴ Two milliequivalents of NO_3^- = 2 x 62 = 124 mg of NO_3^-
∴ Amount of NO_3^- required for replacing two milliequivalents of SO_4^{2-} is 124 mg of NO_3^-

---

798*. Note that 1 ha = 10,000 m^2, depth of furrow slice = 15 cm and B.D. of the soil = 1.335 g/cm^3
One ha = 10,000 m^2 = 10,000 x 10000 cm^2
Volume of ha furrow slice soil = 10,000 x 10000 x 15 cm^3
Since B.D. of the soil is 1.335 g/cm^3, the mass of ha furrow slice soil = 10,000 x 10000 x 15 x 1.335 g
= 10,000 x 10000 x 15 x 1.335 x 10^{-3} kg = 2 x 10^6 kg = 2 million kg = 2 x 10^9 g
Now, $\frac{1\,g}{2 \times 10^9\,g}$ = 0.5 x 10^{-9} = half of a billionth

---

820*. Conversion from ^{0}C to ^{0}F = [32 + temperature in ^{0}C x (9/5)]^{0}F
∴ 25^{0}C = [32 + 25 x (9/5)]^{0}F = 77^0 F

830*. Conversion from ^{0}F to ^{0}C = [(temperature in ^{0}F -32) x (5/9)]
∴ 105^{0}F = [(105-32) x (5/9)]^0 C = [73 x (5/9)]^{0}F = 40.55^0 C

---

840. Porosity, B.D. and P.D. are interrelated as,
% Pore space = 100 [1- (B.D./P.D.)]
In the present case, B.D. and P.D. are 1.4 and 2.8 g/cm^3
Therefore, Per cent pore space =

50 = [1 - (B.D./P.D.)] x 100 =
[1 - (B.D./2.8)] x 100 =
Or, 140 = 280 – 100 B.D.
B. D. = 1.4 g/cm^3
B. D. is increased from 0.7 to 1.4 g/cm^3 or doubled.

844*. To start with, find out the solubility in mole/L (S) from the K_{sp} and then the same in g/L from 'S x molecular weight'. Note that the molecular weights of $Ca(OH)_2$, $CaCO_3$ and $CaSO_4$ are 74, 100 and 136, respectively.

For $Ca(OH)_2$ – $K_{sp} = [Ca^{2+}]\,[OH^-]\,[OH^-] = [Ca^{2+}]\,[OH^-]^2 =$
$(S) \times (2S)^2 = 4\,S^3$.

$\therefore S = \sqrt[3]{\frac{K_{sp}}{4}} = [\frac{5.5 \times 10^{-6}}{4}]^{1/3} = 1.111 \times 10^{-2} =$

$1.111 \times 10^{-2} \times 74 =$ 0.822 g/L

For $CaCO_3$ – $K_{sp} = [Ca^{2+}]\,[CO_3^{2-}] = S \times S = S^2$

$\therefore S = \sqrt{K_{sp}} = \sqrt{4.8 \times 10^{-9}} = 6.928 \times 10^{-5} =$

$6.928 \times 10^{-5} \times 100 =$ 0.00693 g/L

For $CaSO_4$ – $K_{sp} = [Ca^{2+}]\,[SO_4^{2-}] = S \times S = S^2$

$\therefore S = \sqrt{K_{sp}} = \sqrt{1.2 \times 10^{-6}} = 1.095 \times 10^{-3} =$

$1.095 \times 10^{-3} \times 136 =$ 0.149 g/L

Therefore, the solubility of $CaCO_3$ is the least of the three salts followed by $CaSO_4$ and $Ca(OH)_2$.

---

848. For $H_2C_2O_4$ (oxalic acid), the element undergoing change in Oxidation Number is C.

The O.N. of C (C_2) in the reactant $H_2C_2O_4$ is +6 (+2 +6 -8). The O.N. of C in the product CO_2 is +4. So, for two C atoms, the change in O.N. is +8. The net change in O.N. for C for 1 molecule of $H_2C_2O_4$ is therefore +6 to +8 = 2

Therefore, the equivalent weight of $H_2C_2O_4$ = M.W./2= 90/2= 45

852*. Titre value is 5.2 ml 0.02 N $AgNO_3$ for 10 ml soil water extract

5.2 ml 0.02 N $AgNO_3$ = 5.2 x 0.02 ml 1 N $AgNO_3$ ≡ 5.2 x 0.02 ml 1 N chloride =

5.2 x 0.02 x 0.03546 g chloride for 10 ml SWE (1 ml 1 N chloride = 35.46/1000 = 0.03546 g chloride)

∴ 1 L (1000 ml) SWE contains 5.2 x 0.02 x 0.03546 x (1000/10) = 0.36878 g chloride

---

856*. Porosity, B.D. and P.D. are interrelated as,

% Pore space = $100\,(1-\frac{B.D.}{P.D.})$

In the present case, B.D. and P.D. are 1.4 and 2.8 g/cm^3

∴ % pore space is = $100\,(1-\frac{1.4}{2.8}) = 50$

Now, B.D. is decreased from 1.4 to 0.7 g/cm^3

Inserting the changed value of B.D. and the original value of P.D., we have

% pore space = $100\,(1-\frac{B.D.}{P.D.}) = 100\,(1-\frac{0.7}{2.8}) = 75$

∴ % pore space of the soil is increased from 50 to 75 i. e. increased by 50% of its original value.

---

875. For $Fe_2(SO_4)_3$, the element undergoing change in oxidation number is Fe.

The O.N. of Fe (Fe_2) in the reactant $Fe_2(SO_4)_3$ is +6. The O.N. of Fe in the product $FeSO_4$ is +2. So, for two Fe atoms, the change in O.N. is +4. The net change in O.N. for Fe for 1 molecule of $Fe_2(SO_4)_3$ is therefore

+6 to +4 = 2

Therefore, the equivalent weight of $Fe_2(SO_4)_3$ = M.W./2= 399.9/2= 199.95

876*. H$^+$ concentration of a solution of pH 6 = 10^{-6} g/L, while that of a solution of pH 7 = 10^{-7} g/L.

This implies that an increase of pH of a solution by 1 unit (say, 6 to 7) requires 10 fold dilution. Therefore, 1 ml solution for 1 unit rise in pH requires 9 ml water (1 + 9 = 10).

∴ 100 ml solution of pH 6 requires 900 ml water for raising its pH to 7.

903*. One ha soil requires 300 kg SSP (16% P_2O_5). So, 0.5 ha soil requires 150 kg SSP.

150 kg SSP = 150 x (16/100) = 24 kg P_2O_5

∴ In the present situation, due to nonavailability of SSP, the farmer has to apply 24 kg P_2O_5 as TSP (46% P_2O_5) to his 0.5 ha field.

24 kg P_2O_5 = 24 x (100/46) = 52.17 kg TSP

∴ Due to nonavailability of SSP, the farmer has to apply 52 kg TSP to his 0.5 ha field.

907*. Note that Lime Requirement of the acid soil is 5.5 tons/ha, field size is 0.5 ha, molecular weight of limestone ($CaCO_3$) is 100, and molecular weight of burnt lime (CaO) is 56.
∴ 0.5 ha soil requires 2.75 tons limestone. And 100 parts of limestone are equal to 56 parts burnt lime.
∴ 2.75 tons limestone = 2.75 x (56/100) = 1.54 tons of burnt lime.
∴ Due to nonavailability of limestone, the farmer has to apply 1.54 tons of burnt lime to his 0.5 ha field.

930*. The equivalent weights of H^+, Na^+, K^+, Ca^{2+}, and Mg^{2+} are 1, 23, 39, 20, and 12, respectively.
∴ 3, 69, 39, 150, and 18 mg/100 g soil of H^+, Na^+, K^+, Ca^{2+}, and Mg^{2+} are equal to 3, 3, 1, 7.5 and 1.5 m. e./100 g soil of the respective ions.
∴ CEC of the soil is 3+3+1+7.5+1.5 = 16. e./100 g soil.
Now, to check the soil to satisfy the requirement of sodic or alkali soil, we have to find out ESP of the soil.
ESP of the soil is (3/16) x 100 = 18.75
Note that, ESP is > 15. So, the soil meets the requirement of an alkali or a sodic soil.

932*. Organic matter content of the soil is 1.035%, while C:N ratio of the soil is 10 to 15:1 (10:1 to 15:1).
Organic matter to organic C conversion factor is 0.58.
∴ Organic C content of the soil is 1.035 x 0.58% = 0.600%
Since, C:N ratio of the soil is to 10:1 to 15:1, N content of the soil is 0.600/15 = 0.040% to 0.600/10 = 0.060%.
Or, N content of the soil is 0.004 to 0.060%

939*. N content of A.S. is 20.6%; N and P_2O_5 contents of DAP are 18% and 46%; P_2O_5 content of SSP is 16%.
Amount of DAP = 150 kg
150 kg DAP = 150 x (18/100) = 27 kg N and 150 x (46/100) = 69 kg P_2O_5
27 kg N = 27 x (100/20.6) = 131 kg A.S. and 69 kg P_2O_5 = 69 x (100/16) = 431 kg SSP
∴ Due to nonavailability of DAP, the farmer has to apply 131 kg A.S. and 431 kg SSP to his field in place of 150 kg DAP.

951*. K_2O content of MoP and SoP is 60 and 50.

∴ 150 kg MoP = 150 x (60/100) = 90 kg K_2O

And, 90 kg K_2O = 90 x (100/50) = 180 kg SoP

∴ As regards K_2O, 150 kg MoP = 180 kg SoP

954*. Acid equivalent of A.S. is 110 which means that application of 100 kg A.S. requires 110 kg lime for neutralization of the acidity formed.

∴ Acidity produced by the application of 180 kg A.S. requires 180 x (110/100) = 198 kg lime

∴ 198 kg lime is to be used for the neutralization of the acidity produced by 180 kg A.S.

974*. Base equivalent of calcium cyanamide is 63 which means that the application of 100 kg calcium cyanamide adds basicity to the soil equal to that produced by 63 kg lime.

∴ Basicity produced by 175 kg calcium cyanamide = (63/100) x 175 = 110 kg lime

∴ Basicity produced by the application of 175 kg calcium cyanamide is equal to the basicity formed by the addition of 110 kg lime.

1011*. Atomic weights of K and O are 39 and 16, respectively.

∴ 94 parts of K_2O = 78 parts K.

∴ 1 part of K_2O = 78/94 = 0.83 part K.

∴ % K_2O content x 0.83 = % K content.

1026*. Amount of Ca + Mg in 10 ml soil water extract = 0.087 m. e.

10 ml SWE = 1.2 mg Ca = 1.2/20 = 0.06 m. e. Ca

∴ 10 ml SWE = (0.087 – 0.06) m. e. Mg = 0.027 m. e. Mg = 0.027 x 12 mg Mg = 0.324 mg Mg

Note that, 10 ml SWE = 0.5 g soil

∴ 0.5 g soil contains 0.324 mg Mg

∴ 100 g soil contains 0.324 x (100/0.5) mg Mg = 64.8 mg Mg = 0.0648 g Mg

∴ The soil contains 0.0648% Mg

1030*. C:N ratio of the soil is 10:1, while C:P ratio of the soil is 150:1
Assume the soil to contain 1 mg C.
In such a case, the soil contains 1/10 = 0.1 mg N and 1/150 = 0.0066 mg P.
∴ N:P ratio of the soil is 0.1 mg/0.00666 mg = 15 or, 15:1
∴ N:P ratio of the soil is 15:1

1046*. Available N content of the soil = 500 mg/kg soil = 0.5 g/kg soil = 0.0005 kg/kg soil = 0.05 kg/100 kg soil = 0.05%
283 kg urea = $\frac{283 \times 46}{100}$ = 130 kg available N (for 1 ha furrow slice soil = 2.2 x 10^6 kg soil)
130 kg available N for 2.2 x 10^6 kg soil = $\frac{130 \times 100}{2.2 \times 10^6}$ = 0.0059 kg/100 kg soil = 0.0059%
∴ Increased available N content of the soil = 0.05% + 0.0059% = 0.0559%

II. Filling the Blanks with Appropriate Answers

Fill in the blanks with right answer/s (* solutions for the Questions with * mark are worked out in brief at the end of the answers)

1. eluviation	**2.** chlorite	**3*.** 0.5	**4.** Suborder
5. any two from: temperature, water, ice, wind, wave	**6.** oxygen	**7.** g/cm^2/min	**8.** zinc (Zn)
9. lithosphere	**10*.** 0.333	**11.** ammonium carbamate	**12.** Millets
13. Arnon and Stout	**14.** Biomass	**15.** Catena	**16.** Schoonover
17. Anthropic	**18.** Subgroup	**19.** naphtha	**20.** urea
21. Xeric	**22.** profile	**23.** immobilization	**24*.** 13.33
25. (atmospheric) CO_2	**26.** Lime Requirement	**27.** ammonium sulphate	**28.** silicon
29. Soils	**30.** magnesium; 1:2	**31.** mineralization	**32.** potassium and aluminium
33*. 85.71	**34.** hydration	**35.** twelve (12)	**36.** phase

37. laterization	**38.** Hypergelic	**39.** Epipedons	**40*.** 31.9; 44.1
41. Penning	**42.** Edaphology	**43.** NH_4^+	**44.** saltation
45*. 9.4	**46.** 10 ha	**47.** Petrology	**48.** m.e./L
49. Fragipan	**50*.** 120 ml	**51.** Al^{3+}	**52.** Gelisol
53. Albedo	**54.** Inceptisol and Entisol	**55.** exfoliation	**56.** silica
57. earth's crust	**58*.** 18 m. e./100 g soil	**59.** dispersion medium	**60.** dolomite
61. ammonium chloride	**62*.** 21.6	**63*.** sulphuric	**64.** electrical conductivity
65. Endopedon	**66.** nitrogenase	**67.** guano	**68.** furrow slice
69. potassium schoenite	**70.** 6 (six)	**71.** Beer's	**72.** mass flow, root interception
73. monocalcium phosphate and gypsum [$Ca(H_2PO_4)_2$ and $CaSO_4$]	**74.** texture	**75.** spherical	**76.** biuret

77. scarification	**78.** natural gas	**79.** nitrophosphates	**80.** isotopes
81. plasticity	**82.** Hyperthermic	**83.** leghemoglobin	**84.** H^+
85. straight	**86.** ammonium polyphosphate	**87.** Schofield	**88*.** 151.85
89. molten lava/magma	**90.** diammonium phosphate (DAP)	**91*.** 0.5	**92.** limonite
93*. 0.680 g	**94.** mulch	**95.** Great Group	**96.** iron oxides
97. green	**98.** sodium chloride	**99.** biosphere	**100.** alluvial
101. chemoautotrophs	**102.** molybdenum (Mo)	**103*.** 233 kg A.S. & 240 kg neem cake	**104.** endopedons
105. Andisol	**106.** Meghdoot	**107.** triple super phosphate	**108.** methane

109. sulphuric and nitric acid (H_2SO_4 and HNO_3)	**110.** 10 to 15:1	**111.** Psychrophiles	**112.** $NO_2^- \rightarrow NO_3^-$ (nitrite to nitrate)
113. NO_3^- (nitrate)	**114.** slickensides	**115.** illite	**116.** sulphide
117. Podzolisation	**118.** copper	**119.** nitrogen	**120.** nickel (Ni)
121. saline sodic (saline alkali)	**122.** 2.53	**123.** basic slag	**124*.** 82.8
125. hydrogen	**126.** ^{241}Am; Beryllium (Be)	**127.** cohesion and adhesion	**128.** Ca^{2+} (calcium)
129. eutrophication	**130.** Landfills	**131.** sodium nitrate ($NaNO_3$)	**132.** Volts (V) or milli Volts (mV)
133. sodium chloride (NaCl)	**134.** wavelength (λ)	**135*.** 0.003932	**136.** oxygen
137. lignins	**138*.** 1100	**139.** hollow cathode lamp	**140.** Biochar
141. ammonium acetate (CH_3COONH_4)	**142.** sodium hexameta phosphate	**143*.** 0.045	**144.** magnesium (Mg)
145. poudrette	**146.** water	**147*.** 0.768	**148.** $(HPO_4)^-$

149. gypsum	**150.** $K_2Cr_2O_7$ (potassium dichromate)	**151.** regolith	**152.** methane (CH_4)
153. single super phosphate	**154.** phase	**155.** nitrate (NO_3-)	**156.** gleyed
157. sulphuric (H_2SO_4)	**158.** scanner	**159.** 2000	**160.** ustic

Solutions for the Questions with * Mark

3*. The equivalent weights of Ca^{2+}, Mg^{2+}, CO_3^{2-} and HCO_3^- are 20, 12, 30 and 61, respectively.

∴ 80, 24, 165 and 61 mg/L of Ca^{2+}, Mg^{2+}, CO_3^{2-} and HCO_3^- are equal to 4, 2, 5.5 and 1 m.e./L of the respective ions.

∴ RSC = ([CO_3^{2-}] + [HCO_3^-]) – ([Ca^{2+}] + [Mg^{2+}]) = (5.5+1) – (4+2) = 6.5 - 6 = 0.5 m.e./L

10*. Atomic weights of O and S are 16 and 32, respectively.

∴ 96 parts of SO_4^{2-} = 32 parts S.

∴ 1 part of SO_4^{2-} = 32/96 = 0.333 part S.

∴ % SO_4^{2-} content x 0.333 = % S content.

24*. The equivalent weights of H^+, Na^+, K^+, Ca^{2+}, and Mg^{2+} are 1, 23, 39, 20, and 12, respectively.

∴ 3, 46, 39, 150, and 18 mg/100 g soil of H^+, Na^+, K^+, Ca^{2+}, and Mg^{2+} are equal to 3, 2, 1, 7.5 and 1.5 m. e./100 g soil of the respective ions.

∴ CEC of the soil is 3+2+1+7.5+1.5 = 15 m. e./100 g soil.

∴ %Na^+ saturation of the soil (ESP) is (2/15) x 100 = 13.33%

33*. The equivalent weights of H^+, Na^+, K^+, Ca^{2+}, and Mg^{2+} are 1, 23, 39, 20, and 12, respectively.

∴ 2, 23, 58.5, 160, and 18 mg/100 g soil of H^+, Na^+, K^+, Ca^{2+}, and Mg^{2+} are equal to 2, 1, 1.5, 8 and 1.5 m. e./100 g soil of the respective ions.

∴ CEC of the soil is 2+1+1.5+8+1.5 = 14 m. e./100 g soil.
∴ PBS of the soil is [(1+1.5+8+1.5)]/14 x 100 = 12/14 = 85.71%.

40*. Mass of moisture in the soil = mass of moist soil – mass of dry soil = 126.4 g – 95.8 g = 30.6 g

Therefore, 95.8 g soil contains 30.6 g moisture.

Therefore, gravimetric % moisture content of the soil is [(30.6 g/95.8 g) x100] = 31.9

B.D. of the soil is 1.38 g/cm^3. Therefore, volume of the soil is (95.8/1.38) = 69.4 cm^3

Therefore, 69.4 cm^3 soil contains 30.6 cm^3 moisture (30.6 g water = 30.6 cm^3 water)

Therefore, volumetric % moisture content of the soil is (30.6/69.4) x100 = 44.1

45*. Titre value is 4.7 ml 0.1 N H_2SO_4 = 4.7 x 0.1 = 0.47 ml 1 N H_2SO_4

0.47 ml 1 N H_2SO_4 = 0.47 m. e.

∴ 0.47 m. e. is present in 5 g soil.

∴ 100 g soil contains 0.47 x (100/5) = 9.4 m. e.

∴ The CEC of the soil is 9.4 m. e./100 g soil.

50*. CEC of montmorillonite is 120 m.e./100 g which implies that 100 g H^+-montmorillonite contains 120 m.e. H^+.

Therefore, 5 g H^+-montmorillonite contains (120 x 5/100) = 6 m.e. H^+

6 m.e. H^+ = 6 ml 1 N H^+ = 6 ml 1 N KOH = 120 ml 0.05 N KOH

Therefore, 120 ml 0.05 N KOH is required for the said titration.

58*. The equivalent weights of H^+, Na^+, K^+, Ca^{2+}, and Mg^{2+} are 1, 23, 39, 20, and 12, respectively.

∴ 3, 46, 39, 190, and 30 mg/100 g soil of H^+, Na^+, K^+, Ca^{2+}, and Mg^{2+} are equal to 3, 2, 1, 9.5 and 2.5 m. e./100 g soil of the respective ions.

∴ CEC of the soil is 3+2+1+9.5+2.5 = 18 m. e./100 g soil.

62*. Rate of application of A.S. is 180 kg/ha. So, 0.5 ha land requires 90 kg A.S.

100 kg A.S. contains 24 kg S.

∴ 90 kg A.S. adds (24/100) x 90 = 21.6 kg S to the field of 0.5 ha.

63*. Note that 2.85 ml concentrated acid forms 1 L 0.1 N solution of the acid.
To find out the normality of the concentrated acid, use $V_1 N_1 = V_2 N_2$ formula, where
V_1 = volume of the concentrated acid used (= 2.85 ml), N_1 = normality of the concentrated acid (= ?), V_2 = volume of the prepared acid solution (= 1 L = 1000 ml) and N_2 = normality of the prepared acid solution (= 0.1).

$$\therefore N_1 = \frac{V_2 \times N_2}{V_1} = \frac{1000 \text{ ml} \times 0.1 \text{ N}}{2.85 \text{ ml}} = 35 \text{ N}$$

Recall that the normality of concentrated sulphuric acid is 34 to 36 N.
So, the analyst used concentrated sulphuric acid.

88*. From the reaction, we find that 1 mole of $K_2Cr_2O_7$ = 6 equivalents of it. It therefore implies from the reaction, 6 moles of $FeSO_4$ = 6 equivalents of $FeSO_4$.
∴ Equivalent weight of $FeSO_4$ is its molecular weight $FeSO_4$ i. e. 151.85.

91*. Concentrations of Na^+, Ca^{2+} and Mg^{2+} of the water source are 1, 6 and 2 m. e. of the respective ions per L.
Therefore, inserting the values of Na^+, Ca^{2+}, and Mg^{2+} in the expression for SAR, we have

$$SAR = \frac{[Na^+]}{\sqrt{\frac{[Ca^{2+}] + [Mg^{2+}]}{2}}} = \frac{1}{\sqrt{\frac{6+2}{2}}} = ½ = 0.5$$

93*. One L 500 ppm Fe solution requires 500 mg Fe.
∴ 0.5 L 500 ppm Fe solution requires 250 mg Fe.
Molecular weight of $FeSO_4$ is 151.85 which implies that 151.85 parts of $FeSO_4$ contain 55.85 parts of Fe.
∴ 250 mg Fe = (151.85/55.85) x 250 = 680 mg $FeSO_4$ = 0.680 g $FeSO_4$.
∴ 0.680 g accurately weighed $FeSO_4$ is to be dissolved in water and diluted to 500 ml for preparing 0.5 L 500 ppm Fe solution.

103*. Rate of application of N is 120 kg/ha. Field size is 0.5 ha.
∴ 0.5 L ha land requires 60 kg N in 4:1 ratio from A.S. (20.6% N) and neem cake (5% N).
∴ 0.5 L ha land requires (4/5) x 60 = 48 kg N from A.S. and (1/5) x 60 = 12 kg N from neem cake.
48 kg N = (100/20.6) x 48 = 233 kg A.S.
12 kg N = (100/5) x 12 = 240 kg neem cake

∴ 0.5 L ha field requires 233 kg A.S. and 240 kg neem cake for applying N @ 120 kg/ha with the condition of adding N from A.S. and neem cake in 4:1 ratio.

124*. CEC of the soil is 24 m. e./100 g soil
Note the ESP of alkali soils must be > 15.
∴ Exchangeable sodium content of the alkali soil must be a minimum of $\frac{24 \times 15}{100} = 3.6$ m. e./100 g soil
3.6 m. e. exchangeable sodium /100 g soil = 3.6 x 23 = 82.8 mg exchangeable sodium /100 g soil (note that 23 is the equivalent weight of sodium)

135*. 10 ppm Cu solution contains 10 mg Cu in 1 L solution.
∴ 100 ml 10 ppm Cu solution contains 1 mg Cu.
From the atomic and molecular weights of Cu and $CuSO_4.5\ H_2O$, we have,
63.5 mg Cu = 249.7 mg $CuSO_4.5\ H_2O$.
∴ 1 mg Cu = 249.7/63.5 = 3.932 mg $CuSO_4.5\ H_2O$.
∴ 100 ml 10 ppm Cu solution contains 3.932 mg $CuSO_4.5\ H_2O$ = 0.003932 g $CuSO_4.5\ H_2O$.

138*. Volume of a cylinder is given by the formula $\pi r^2 h$.
In the present case, r = 5 cm and h = 14 cm.
∴ Volume of the cylindrical core sampler = $\pi r^2 h = (22/7) \times 5^2 \times 14 = 1100$ cm^3

143*. 30 mg Mg^{2+} = 30/12 = 2.5 m.e. $Mg^{2+} \equiv$ 2.5 m.e. NH_4^+ = 2.5 x 18 = 45 mg NH_4^+ = 0.045 g NH_4^+

147*. Titre value is 3.2 ml 0.02 N EDTA = 3.2 x 0.02 ml 1 N EDTA = 0.064 ml 1 N EDTA ≡ 0.064 ml 1 N Mg^{2+}
The equivalent weight of Mg^{2+} is 12.
∴1 ml 1 N Mg^{2+} is 0.012 g = 12 mg Mg^{2+}
∴ 0.064 ml 1 N Mg^{2+} = 0.064 x 12 mg Mg^{2+} = 0.768 mg Mg^{2+}
∴ A titre value is 3.2 ml 0.02 N EDTA corresponds to 0.768 mg Mg

III. Formulae / Compositions of Minerals/Fertilizers

Furnish the formulae OR chemical compositions of the following minerals/fertilizers

1.	Actinolite	$Ca_2(Mg.Fe)_5Si_8O_{22}(OH)_2$
2.	Albite	$NaAlSi_3O_8$
3.	Ammonium carbamate	$H_2NCOONH_4$
4.	Anatase	TiO_2
5.	Andalusite	Al_2SiO_5
6.	Anhydrite	$CaSO_4$
7.	Anorthite	$CaAl_2Si_2O_8$
8.	Apatite	$Ca_{10}(PO_4)_6(OH.F.Cl)_2$
9.	Azurite	$2\ CuCO_3.Cu(OH)_2$
10.	Barite	$BaSO_4$
11.	Bauxite	$Al_2O_3.2\ H_2O$
12.	Beryl	$3\ BeO.Al_2O_3.6\ SiO_2$
13.	Biotite	$(K_2Al_2Si_6)(Fe.Mg)_6O_{20}(OH)_4$
14.	Biuret	$H_2NCONHCONH_2$
15.	Borax	$Na_2B_4O_7.10\ H_2O$
16.	Bornite	Cu_5FeS_4
17.	Brucite	$Mg(OH)_2$
18.	Calamine	$ZnCO_3$
19.	Calcite	$CaCO_3$
20.	Carnallite	$KCl.MgCl_2.6\ H_2O$
21.	Chalcopyrite (copper pyrite)	$CuFeS_2$

22.	Chalcocite	Cu_2S
23.	Chile salt peter	$NaNO_3$
24.	Colemanite	$Ca_2B_6O_{11}.5\ H_2O$
25.	Corundum	Al_2O_3
26.	Covellite	CuS
27.	Cryolite	Na_3AlF_6
28.	Dicalcium phosphate	$CaHPO_4$
29.	Dolomite	$CaCO_3.MgCO_3$ or $CaMg(CO_3)_2$
30.	Emerald	$Be_3Al_2(SiO_3)_6$
31.	Enstatite	$Mg_2Si_2O_6$
32.	Epidote	$Ca_2Fe^{(III)}Al_2Si_3O_{12}OH$
33.	Epsomite	$MgSO_4.7\ H_2O$
34.	Fayalite	Fe_2SiO_4
35.	Ferrosilite	$Fe_2Si_2O_6$
36.	Fluorapatite	$Ca_{10}(PO_4)_6F_2$ or, $3\ Ca_3(PO_4)_2.CaF_2$
37.	Fluorspar	CaF_2
38.	Forsterite	Mg_2SiO_4
39.	Francolite	$Ca_{10}(PO_4)_6CO_3$
40.	Friedel salt	$3\ CaO.Al_2O_3.CaCl_2.10\ H_2O$
41.	Galena	PbS
42.	Garnet	$Fe_3Al_2(SiO_4)_3$ (almandine)
43.	Gibbsite	$Al(OH)_3$
44.	Glauber's salt	$Na_2SO_4.10\ H_2O$
45.	Goethite	$FeO.OH$

46.	Gypsum	$CaSO_4.2\ H_2O$
47.	Halite	$NaCl$
48.	Hausmannite	$Mn_3O_4\ (Mn^{II}Mn^{III}{}_2O_4)$
49.	Hematite	Fe_2O_3
50.	Ilmenite	$FeTiO_3$
51.	Inderite	$MgB_3O_3(OH)_5.5\ H_2O$
52.	Kainite	$KCl.MgSO_4.3\ H_2O$
53.	Kaolinite	$Si_4Al_4O_{10}(OH)_8$
54.	Kieserite	$MgSO_4.H_2O$
55.	Kyanite	Al_2SiO_5
56.	Langbeinite	$K_2SO_4.2\ MgSO_4$
57.	Leucite	$K_2Al_2Si_4O_{12}$
58.	Limonite	$2\ Fe_2O_3.3\ H_2O$
59.	Magnesite	$MgCO_3$
60.	Magnetite	Fe_3O_4 or $FeO.Fe_2O_3$
61.	Malachite	$CuCO_3.Cu(OH)_2$
62.	Manganite	$MnOOH$
63.	Mohrite	$(NH_4)_2SO_4.FeSO_4.6\ H_2O$
64.	Molybdenite	MoS_2
65.	Muscovite	$(K_2Al_2Si_6)Al_4O_{20}(OH)_4$
66.	Nepheline	$NaAlSiO_4$
67.	Nitre	KNO_3
68.	Orthoclase	$KAlSi_3O_8$
69.	Oxyapatite	$Ca_{10}(PO_4)_6O$

70.	Perovskite	$CaTiO_3$
71.	Polyhalite	$K_2SO_4.MgSO_4.2\ CaSO_4.2\ H_2O$
72.	Potassium schoenite	$K_2SO_4.MgSO_4.6\ H_2O$
73.	Powellite	$CaMoO_4$
74.	Pyrite	FeS_2
75.	Pyrolusite	MnO_2
76.	Quartz	SiO_2
77.	Rhodochrosite	$MnCO_3$
78.	Rutile	TiO_2
79.	Sanidine	$KAlSi_3O_8$
80.	Serpentine	$Mg_6Si_4O_{10}(OH)_8$
81.	Siderite	$FeCO_3$
82.	Single super phosphate	$Ca(H_2PO_4)_2.CaSO_4$
83.	Smithsonite	$ZnCO_3$
84.	Sphalerite	ZnS
85.	Spinel	$MgAl_2O_4$
86.	Stengite	$FePO_4.2\ H_2O$
87.	Super phosphoric acid	$H_4P_2O_7$
88.	Sylvinite	$KCl.NaCl$
89.	Sylvite	KCl
90.	Talc	$Mg_6Si_8O_{20}(OH)_4$
91.	Thiourea	$CS(NH_2)_2$
92.	Titanite/Sphene	$CaO.TiO_2.SiO_2$
93.	Topaz	$Al_2SiO_4(OH.F)_2$

94.	Triple super phosphate	$Ca(H_2PO_4)_2$
95.	Variscite	$AlPO_4.2\ H_2O$
96.	Vivianite	$Fe_3(PO_4)_2.8\ H_2O$
97.	Wollastonite	$CaSiO_3$
98.	Wulfenite	$PbMoO_4$
99.	Zinc blende	ZnS
100.	Zircon	$ZrSiO_4$

IV. Contributions of Scientists in Soil Science

Mention the major contribution/s of the following scientists in the field of Soil Research/Soil Science

1. J. Tyndall – Optical properties of colloidal solutions known as Tyndall effect.
2. C. F. Marbut – Pedology, and Zonal, Intrazonal and Azonal systems of soil classification.
3. G. W. Robinson – Early soil scientist known especially for a method for Particle Size Analysis of soils (Robinson's Pipette method).
4. A. H. Munsell – Colour chart for assigning the characteristics of soil colour viz. hue, value and chroma of soil colour.
5. E. W. Hilgard – Pedology, soil formation, and effects of climate, vegetation and rocks in soil formation.
6. D. I. Arnon and P. R. Stout – Established the criteria for the essentiality of plant nutrients.
7. Dr C. N. Acharya – Storage, collection and preparation of farm yard manure FYM, and worked out an effective method of composting known as Bangalore method of composting.
8. Guy D. Smith – Soil survey, soil taxonomy, soil classification, pedology, soil genesis, and soil conservation.
9. A. Howard and Y. D. Wad – Collection and storage of farm yard manure FYM and worked out an effective method of composting known as Indore method of composting.
10. Hans Jenny – Pedology, soil formation, and factors of soil formation.
11. R. K. Schofield – Soil physics, soil chemistry, concepts of soil moisture tension in terms of pF, Ratio law, and ion exchange selectivity.
12. Thomas Graham – Introduced the term colloid, chemistry of colloids, and the two phases of a colloidal system.
13. H. P. J. Darcy – Soil physics and hydraulic flow of a fluid such as water in a porous medium like soil (Darcy's law).
14. Friedrich Wohler – First synthesized urea (an organic compound) in the laboratory (outside animal kidney) by heating ammonium cyanate

which disproved the 'Theory of Vitalism' for the synthesis of organic compounds

15. Friedrich Mohs – Chemist and minerologist and introduced a scale for expressing the hardness of minerals, known as Mohs scale.

16. H. E. Shoemaker – Soil chemist especially known for devising a method for determining the Lime Requirement (LR) of acid soils.

17. J. H. Gilbert – Early soil scientist especially known for contribution to establishing Rothamsted Experimental Station in U.K. along with J. B. Lawes, a pioneering soil scientist in soil fertility studies.

18. J. B. Lawes – Early soil chemist and known for contribution to establishing Rothamsted Experimental Station in U.K. along with J. H. Gilbert, a pioneering soil scientist in soil fertility studies and, manufactured super phosphate by treating coprolites with sulphuric acid which was the stepping stone for the commencement of production of commercial chemical fertilizers.

19. J. Samuel Joffe – Early soil chemist especially known for contribution to the study of pedology, soil formation and categorization of the factors of soil formation into Active and Passive factors.

20. G. G. Stokes – Quantifying the terminal velocity of a falling particle in a viscous medium (Stokes' law) and soil mechanics.

21. V. V. Dokuchaev – Early soil scientist particularly known for his works in pedology, often considered as the Father of Pedology, soil formation and Zonal, Intrazonal and Azonal system of soil classification.

22. Robert Brown – Botanist and colloidal chemist, known for studying the mechanical behaviour of the particles of the dispersed phase of a colloidal system in a dispersion medium (Brownian movement).

23. J. W. Leather – Pioneering researcher in soil science and agricultural chemistry, often regarded as the Father of Soil Science in India.

24. T. C. Chamberlin and F. R. Moulton – Soil geology and formation of the planet earth, much known for his Planetesimal hypothesis, explaining the formation of the earth.

25. H. B. Hutchinson and E. H. Richards – Collection and storage of farm yard manure FYM and worked out an effective method of composting known as ADCO (Agricultural Development Company) method of composting.

26. H. Lundegardh – Soil scientist with remarkable contribution in the study of absorption of plant nutrients and 'cytochrome pump' or electron theory for the uptake of plant nutrients from soil.

27. E. R. Beaufils – Introduced the Diagnosis Recommendation Integrated System (DRIS) concept of applying plant nutrients as fertilizers based on the results of plant analysis.

28. S. P. L. Sorenson – Chemist known for introducing the concept of pH in terms of a numerical scale running from 0 to 14 for expressing the degree of acidity and alkalinity of a solution.

29. Justus von Liebig – Early agricultural chemist who for the first time disproved the 'Humus Theory of Plant Nutrition', demonstrated the increase in crop yields by adding minerals and inorganic materials to soils, and established the role of certain essential factors that limit plant growth and yield, known as 'von Leibig's Law of Minimum'.

30. Thomas Way – Soil chemist who first reported the cation exchange behaviour of surface active substances in equilibrium with the surrounding medium and also, the cation exchange capacity of soils.

31. Jean Baptiste Boussingault – Early agricultural chemist who proved that for plant nutrition, air and water are the primary sources of carbon, hydrogen, and oxygen in plants. Much revered Jean Baptiste Boussingault is often considered as the 'Father of Field Plot Technique' in agriculture.

32. Adolf Fick – Contributed to fundamental studies in physics and especially known for his works on the rate of movement of solutes or particles (diffusion) from their higher to lower concentration regions (Fick's law of Diffusion).

33. M. W. Beijerinck – Microbiologist and reported a number of species of soil bacteria.

34. W. R. Schoonover – Soil chemist especially known for devising a method for determining the Gypsum Requirement (GR) of alkali soils.

35. A. M. Atterberg – Soil physicist who contributed to basic studies in soil physics, retention of soil moisture and introduction of Soil Consistence Limits (Atterberg's limits).

36. L. A. Richards – Contributed to soil physics, operation of tensiometer, soil-water-plant relationship and studies on soil salinity.

37. S. Winogradsky – Soil microbiologist who reported many new species of soil bacteria, and nitrogen transformations in soil.

38. Theodore de Schwann – Early soil scientist and agricultural chemist known for his works on plant nutrition and experiments on plants deriving carbon (C) from air and not from soil.

39. Fritz Haber and Carl Bosch – Industrial manufacture of ammonia from elemental nitrogen and elemental hydrogen.

40. Fowler and Ridge – Demonstrated an effective method of composting from farm wastes known as Activated Composting Process.

41. C. C. Nikiforoff – Pedogenesis, weathering of soils,and soil physics.

42. Konstantin Glinka – Explanation of the concept for geographical distribution of soils, pedogenesis, soil formation processes, podzolization, formation of salt rich soils.

43. R. B. Cate and L. A. Nelson – Known for introducing the concept of 'Critical Levels of Plant Nutrients' based on soil test results and 'Deficiency' and 'Sufficiency' ranges of plant nutrients in soils for applying fertilizers to soils.

44. George N. Coffey – Known for pioneering works in soil classification, soil survey, soil mapping (cartography), and first published the Soil Map of USDA way back in 1912.

45. Friedrich Albert Fallou – Soil geology, pedology (introduced the term too), soil geography, authored a fundamental book on pedology, and considered as the German Founder of Modern Soil Science.

46. Milton Whitney – Contributed to the development of Soil Survey and championed the cause for the significance of soil survey in the development of agriculture.

47. C. G. Hopkins – Pioneering research work in soil fertility and developed effective soil management system based on limestone, rock phosphate, and legumes.

48. R. Warrington – Agricultural chemist, soil microbiologist and biological nitrogen fixation.

V. Definition Explanations / Lucid Notes

Define or explain or write a lucid note on the followings

1. Acid – An acid is defined as a substance which ionizes in aqueous solutions to produce H^+ ions or can donate protons (H^+) to another substance or can take up an electron pair to form a covalent bond. Accordingly, HCl, NH_4^+ and $AlCl_3$ are acids.

2. Acid Equivalent of fertilizers – Acid Equivalent (A. E.) of fertilizers is defined as the amount of lime ($CaCO_3$) required for neutralizing the acidity produced through the application of acid forming chemical fertilizers. It is expressed as a number which indicates the number of kgs of $CaCO_3$ required for neutralizing the soil acidity produced by the application of 100 kg of a fertilizer. Acid equivalents of urea, ammonium sulphate and ammonium chloride are 80, 110 and 129 implying that 80, 110 and 129 kg of $CaCO_3$ are to be applied to soils for the neutralization of the soil acidity produced by the application of 100 kg urea, ammonium sulphate and ammonium chloride, respectively.

3. Acid rain – Acid rain is the rain or precipitation (rain, fog, mist, snow etc.) containing nitric, sulphuric, nitrous, and sulphurous acid (the last two are transient acid molecules). These acids are formed in the rain water due to the dissolution of the acidic oxides of nitrogen and sulfur viz. nitrous oxide (NO), nitric oxide (NO_2) and sulfur dioxide (SO_2) present in the atmosphere. These acidic gases are generated through the burning of fossil fuels, automobile exhausts, industrial discharges and many such activities. Acid rains are more dangerous for water bodies like lakes than they are for soils, the reason being the buffering capacity of soils to counter the effects of acid addition to a certain point beyond which the soils too turn acidic and unproductive. Most fish fail to survive at pH below 4.5 and therefore, acid rains practically eliminate the fish and aquatic animal population of lakes. Acid rains cause also corrosion of metals, buildings, bridges, towers, and human made heritage structures made of marble and limestone. The two ways to handle the problems of the occurrence of acid rains are reducing the emission of the acidic oxides (NO, NO_2 and SO_2) into the atmosphere and resort to liming of the water bodies and soils affected by acid rains.

4. Acid sulphate soils – Acid sulphate soils are formed through natural processes in coastal and inland water areas where waterlogged conditions with sufficient organic matter, iron and sulphate are prevalent. These conditions are suitable for the formation of reduced iron sulphides such as pyrite (FeS_2) and iron monosulphide (FeS), the active acid ingredients of acid sulphate soils. Acid sulphate soils are dark brown to black in colour, rich in organic carbon and sandy to clayey in texture. Acid sulphate soils pose no risk and environmental hazard so long as they are left undisturbed; but, when tilled or drained, the reduced iron sulphides get oxidized forming sulphuric acid and the soil pH drops below 4. These soils therefore, pose a significant environmental risk due to low soil pH, and transport of the acidity into the groundwater or surface water systems through leaching. Acid sulphate soils are common in areas in close proximity to estuaries, wetlands, and shallow groundwater table where waterlogged conditions prevail for long periods. In India, large areas of acid sulphate soils are found in coastal areas of Kerala, locally referred as *Kari* soils of Kuttanad.

5. Activated Sludge Process - Activated Sludge Process is a process for making the best manurial product and recovery of reusable and recyclable materials from the waste water of the towns and cities. The sewage through the sewerage system is taken to an isolated spot in the locality where it is processed by a series of sewage tanks fitted with pipes at different levels and compressors to pump in air under high pressure. The system uses sieves for recovery of materials like glass, plastic etc., chemicals like iron salts, and cultures of aerobic microorganisms. The product is activated sludge, a brown colour odour free concentrated organic manure. The process generates a good manure and a large volume of water best suited for irrigation, recovers many valuable materials like plastic, glass and metals, creates substantial income for local bodies, like the municipalities and corporations and prevents the nearby rivers and tanks from getting polluted.

6. Active acidity of soils - Active acidity of a soil is defined as the soil's acidity due to the presence of H^+ and Al^{3+} ions in the soil solution. A pH meter records the active acidity of a soil which is thus the soil's measurable acidity.

7. Adhesion – Adhesion refers to the attraction between two dissimilar particles or substances *i.e.* between soil particles and water molecules. Adhesion involves diverse kinds of bondings

8. Adsorption – Adsorption is defined as a process in which atoms or ions or molecules of a substance (adsorbate) adhere to a surface of the adsorbent. Adsorption, a surface phenomenon, may be physical or chemical and involves *van der Waals* forces, hydrogen bonding and even ionic bonds as the force of binding between the adsorbate and adsorbent.

9. Albedo - The fraction of the incident electromagnetic radiation that is reflected back by the soil or the land surface is termed as the Albedo. Albedo is a function of several soil properties such as soil colour, moisture, roughness or smoothness, and vegetative cover. Albedo of light coloured smooth soil surfaces is higher than that of dark coloured rough soil surfaces.

10. Alkalization of soils - The process in which soils accumulate appreciable quantity of soluble sodium salts in the upper layers is called alkalization (or, solonization) of soils. The soils so formed are called sodic or alkali soils (solonetz). These soils have pH > 8.5 and ESP > 15. High pH of the soils dissolves humus and the black colored solution moves upward giving the soils a distinct black colour. Hence, alkali or sodic soils are also known as black alkali soils.

Sodic soils are of common occurrence in arid and semiarid regions, and in places where soluble carbonates (CO_3^{2-}) and bicarbonates (HCO_3^-) are in abundance. Soluble Ca^{2+} and Mg^{2+} in soil solution react with the CO_3^{2-} to form insoluble $CaCO_3$ and $MgCO_3$. This causes a decline in the concentration of Ca^{2+} and Mg^{2+} in soil solution. To meet this decline, more Ca^{2+} and Mg^{2+} are released from the soil exchange complexes. This in turn causes an increase in the abundance of Na^+ over the soil exchange complexes (ESP >15). High ESP of the soils causes dispersion of the soil aggregates giving unfavourable soil physical properties like poor aeration, low infiltration and impaired internal drainage. These problems lead to a thin friable horizon followed by a dark, and hard impermeable horizon. Clays are deposited in the illuvial horizon giving the soils a typical columnar structure.

11. Alluvial soils – Alluvial soils are of depositional origin and are relatively young soils. In India, the alluvial soils account for nearly 75 million ha (40% land area) and occur extensively in the states of Punjab, Haryana, Uttarakhand, Uttar Pradesh, Bihar, West Bengal, Assam, and Andhra Pradesh, and other coastal regions of India. Alluvial soils are formed on parent materials transported by the river systems viz. Ganga, Brahmaputra, Narmada, Tapti, Cauvery, Mahanadi, Godavari, and Krishna. The rivers carry the products of weathering of rocks constituting the mountains and deposit them along their path as they flow down the slope over the plains to eventually join the sea. Geologically, the alluvial soils are of two types viz. Old and New alluvium referred to as *Bhangar* and *Khadar* in north India. There is also dark coloured silt rich highly fertile Godavari and Krishna deltas in Andhra Pradesh formed from the basaltic materials. Consequent to being young, the alluvial soils do not have well defined soil profiles, but are very productive and support intensive commercial cultivation. Rice, wheat, maize, mustard, and vegetables are some among the crops best suited for alluvial soils. In Soil Taxonomy, the alluvial soils belong to the soil Orders Inceptisol, Entisol, Alfisol and Aridisol.

12. Ammonification – The soil microbiological process in which nitrogen present in complex organic forms like proteins gets converted to NH_3 or NH_4^+ form is called ammonification. Ammonification, an oxidation process, is the first step in the mineralization of organic nitrogen. Ammonifying microorganisms are all aerobic heterotrophs.

13. Anion Exchange Capacity of soils – The capacity of a soil to hold anions and make them available for exchange with other anions is called Anion Exchange Capacity of the soil (AEC). The AEC of soils is expressed in terms of m. e. per 100 g soil or in terms of centi moles of negative charges per kg soil. The latter is written as c mol (e^-) per kg soil. The letter *e*- stands for electron that carries a single unit of negative charge. The numerical value of AEC of soils and soil clays is far less than that of their Cation Exchange Capacity (CEC).

14. Atomic Absorption Spectrophotometry – Atomic Absorption Spectrophotometry is an analytical method for the determination of metallic elements, say zinc or iron, in micro amounts. In atomic absorption spectrophotometry, the sample is first atomized (converted to atoms) by a strong flame, and then electromagnetic radiation of appropriate wavelength emitted from a suitable source (a hollow cathode lamp) is passed through the vaporized sample, at which some of this radiation is absorbed by a certain number of these atoms (say, iron). As the number of these atoms (concentration of iron) increases, the extent of the radiation absorbed also increases, and proportionately to that extent, the intensity of the transmitted radiation decreases. This reduction in the intensity of the transmitted radiation is correlated to the increased absorbance of the sample which in turn is a direct function of the concentration of the element (iron) present in the sample. The concentration of the element (iron) in the sample is then determined by comparing the instrument reading with those of solutions of known concentrations of the same (iron).

15. Atterberg's limits – The state of consistency between solid and liquid range of soils is divided into a number of distinct stages. Water content limits have been imposed for various stages of consistence. These limits called Liquid, Plastic and Shrinkage limits are called Atterberg's limits. Atterberg's Limits are used to draw a relationship between soil moisture content limits and various states of soil consistency.

16. Base – A base is defined as a substance which ionizes in aqueous solutions to produce hydroxyl (OH^-) ions or can accept protons (H^+) from another substance or can furnish an electron pair to form a covalent bond. Accordingly, NaOH, NH_3 and Cl^- are bases.

17. Base Equivalent of fertilizers – Base Equivalent (B. E.) of a fertilizer is a number which shows the basicity produced through the application of 100 kg of a basic fertilizer and is expressed in terms kgs of $CaCO_3$. The base equivalents of calcium cyanamide ($CaCN_2$), $NaNO_3$, KNO_3 and $Ca(NO_3)_2$ are 63, 29, 29 and 21 which imply that the soil application of 100 kg of each of them produces the basicity equal to 63, 29, 29 and 21 kg of $CaCO_3$, respectively.

18. Bedrock – The hard consolidated rock material underlying a soil's parent material with no evidence of weathering is called bedrock. Almost all soils gradually grade to the underlying feebly weathered parent materials and then to the completely unweathered consolidated bed rocks. It however, must not be construed that a soil has necessarily been formed from the underlying bedrock, due mainly to the issue of transportation of soil materials by the agencies of nature.

19. *Bhangar* and *Khadar* – The alluvial soils, formed by deposits by the flowing rivers such as the Ganges in the flood plains of northern India, are locally called *Bhangar* and *Khadar*. Of the two, *Bhangar* soils, coarse in texture and rich in *kankars*, are the older alluvia; they are less fertile as they are above the flood level, and represent upland alluvial tract. By contrast, the *Khadar* soils are rich in fine silt and clay and are located below the flood plains and hence are more fertile.

20. Biological Nitrogen Fixation – Biological nitrogen fixation (BNF) is defined as the conversion of atmospheric elemental nitrogen to simple inorganic forms like ammonia, nitrite, and nitrate or oxides of nitrogen carried out by selected species of bacteria, algae, actinomycetes and even fungi. Nitrogen fixing microorganisms are symbiotic or associative or free living.

21. Biological Oxygen Demand – Biological Oxygen Demand (BOD) refers to the amount of oxygen required by bacteria and other microorganisms to decompose organic matter of a sample, usually water, at a certain temperature under aerobic conditions. The greater the amount of 'organic contaminant load' (sources may be from industries, municipal wastes, waste water from water treatment plants, failing septic tanks, runoff water etc.) of the sample, the greater is the oxygen demand of the microorganisms. Justifiably, BOD is an indicator of the level of contamination of water (of rivers, streams, lakes and ponds) by organic materials. Consequently, the available dissolved oxygen is spent up by the microorganisms, and to that extent, the life of aquatic fauna suffers. BOD is expressed as milli grams of oxygen per L sample during a five day incubation period at 20^0 C. Much surveillance, check and vigil are required when the BOD value of a water body exceeds 9 mg/L water.

22. Biological weathering - Biological weathering refers to the weathering of rocks and minerals performed by biological organisms. The latter include microorganisms like bacteria, fungi, algae, lichens, higher plants, animals like rats and rodents (burrowing animals), other animals and even man. Truly speaking, biological weathering is either physical or chemical weathering, but executed by biological organisms. Biological weathering thus causes both physical disintegration of consolidated rocks and minerals into smaller sizes and alters their chemical composition as well.

23. Bioremediation - Bioremediation is an ecofriendly process of the degradation of toxic pollutants into nontoxic or less toxic substances executed by specific microorganisms through their enzyme systems. Our environment is getting increasingly contaminated by pesticides and other obnoxious substances, and bioremediation is therefore being increasingly exploited to degrade many of these hazardous substances present in soil and water to their nontoxic metabolites by using especially abled genetically engineered bacteria and fungi.

24. Black soils - Black soils or black cotton soils, locally known as *Regurs*, *Murram*, *Mullah*, *Karail*, *Tal* and *Regadi* in various regions, cover an estimated area of 74 million ha in India, and are widespread in the states of Maharashtra, Gujarat, Madhya Pradesh, Chhattisgarh, Andhra Pradesh, Karnataka, Tamil Nadu, and Rajasthan. Black soils are formed from Deccan basalt trap rocks either *in situ* or over the transported parent materials, and are characteristically black coloured due to the presence of titaniferous magnetite and dissolved organic matter. Black soils are fine textured, rich in smectite clays, high in CEC, calcareous, alkaline in pH and highly fertile. Black soils do show deep cracks and slickensides due to swell shrink property and are well suited for cotton, pulses, millets, pea and rice. In Soil Taxonomy, the black soils belong to the soil Orders Vertisol, Inceptisol and Entisol.

25. Brownian movement - The continuous haphazard, erratic and zig zag movement of the dispersed particles of a colloidal system is called the Brownian movement. Brownian movement caused by continuous hits by the molecules of the dispersion medium over the particles of the dispersed phase distinguishes a colloidal solution from a true solution. Brownian movement was first reported by Robert Brown, a botanist, for the movement of pollen grains in water.

26. Buffer - A buffer is a solution that tends to resist changes in pH that are normally affected by the addition of small amounts of acids or bases. Buffers thus helps maintaining a narrow range of pH of a solution. Most buffers are made of a weak acid and its conjugate base.

27. Bulk density of soils - Bulk density of soils is defined as the mass per unit volume of dry soils (volume of soil solids and pore space) and is expressed as g per cm^3 or Mg per m^3. The bulk density of loose and porous soils is lower than those of compact soils because of higher porosity.

28. Carbon Farming - Carbon farming refers to the various farming practices such as growing cover crops, zero tillage or less tillage agriculture, crop rotation, agroforestry, organic farming with no chemical fertilizers etc. that help sequestrating carbon that otherwise ends up as CO_2 in the atmosphere into the soil through plant roots and soil organic matter. Carbon farming helps mitigating the climate change effects and proves very effective in reducing or avoiding greenhouse gas emissions.

29. Carbonation - Reaction of a substance, say insoluble limestone ($CaCO_3$), with dissolved CO_2 to form a new substance, in this case soluble calcium bicarbonate [$Ca(HCO_3)_2$], is referred to as carbonation.

$$CaCO_3 + H_2O + CO_2 \longrightarrow Ca(HCO_3)_2$$

Carbonation is an important reaction in the chemical weathering of rocks and minerals leading to the formation of soils.

30. Catena - A group of soils of about the same age and formed under the same climatic conditions and from the same parent material, but under different relief, is called a Catena. Catena signifies the role of the features of the relief of a land such as elevation, topography, slope and position in the landscape in soil formation and in determining the properties of soils.

31. Cation Exchange Capacity of soils - The capacity of a soil to hold cations and make them available for exchange by other cations is called the Cation Exchange Capacity of the soil (CEC). CEC of soils increases with increase in negative charge of the soil's clays and humus and is expressed in terms of m. e. per 100 g soil or c mol (p^+) per kg soil.

32. c-Axis spacing - c-Axis spacing refers to the vertical distance between two successive repeating units of a layer lattice crystal system such as a silicate clay, say kaolinite. c-Axis spacing is expressed in terms of Å (10^{-10} m) or nm (10^{-9} m). c-Axis spacing of kaolinite, illite and montmorillonite are 7.2, 10 and 14 Å, respectively.

33. Chelate - Chelates are large sized organic ligands with two or more electron rich sites acting as electron donors for binding to metal ions. Chelates can thus hold metal ions viz. Cu^{2+} and Zn^{2+} somewhat tightly and prevent them from being released and are thus of much significance in the sustained release of micronutrients from the soils required for plant nutrition. EDTA and DTPA are two important chelates with six and eight binding sites.

34. Chemical Oxygen Demand - Chemical Oxygen Demand (COD) of a water source, usually a polluted water source such as river, stream, lake, pond, drain, runoff etc., is a measure of the water sample's capacity to consume oxygen during the decomposition of organic matter and the oxidation of simple inorganic entities like NH_4^+, NO_2^- etc. The usual oxidizing agent for the assay of COD is acidified (by H_2SO_4) $K_2Cr_2O_7$. The quantity of $K_2Cr_2O_7$ consumed is expressed as its oxygen equivalence in terms of mg O_2 L^{-1} of water. Since the oxidant ($K_2Cr_2O_7$) is not specific to O_2 consuming chemicals that are organic or inorganic (NH_4^+, NO_2^- etc.), both are included in COD assay. A high COD value of a water source is indicative of its high pollution load and the water may prove damaging to aquatic life forms due to poor dissolved oxygen level. Purity of treated water emanating from industries can thus be monitored by its COD. COD measures the oxygen demand of a representative sample of the whole lot to oxidize all organic contents, while BOD measures just that portion of it which is consumed biologically.

35. Chemical weathering - Chemical weathering refers to the process of weathering of rocks and minerals effecting a change in their chemical composition. Effect of chemical weathering is more pronounced in the humid tropic regions of the earth where the earth's surface remains moist for most part of the year and temperature is also high. Plants, animals and microorganisms also contribute to the chemical weathering of rocks and minerals, either directly or indirectly. Chemical weathering of rocks and minerals involves chemical reactions which include solution, hydrolysis, hydration, oxidation, reduction, and carbonation. Chemical weathering converts primary minerals to secondary minerals like the silicate clays.

36. Chemoautotrophs - Microorganisms which derive their requirements of carbon from CO_2 and the biochemical energy from the oxidation of simple inorganic compounds such as ammonium, sulphide, nitrite etc. are called chemoautotrophs. Nitrifying bacteria such as *Nitrosomonas* and *Nitrobacter* are chemoautotrophs.

37. Chernozems (Chernozem soils) - Chernozems (Russian words *Chernaya* and *Zemlya* meaning black and earth) are characterized by 2 meter deep surface layer rich in humus (upto 16%), nitrogen, phosphorus, lime and having well aggregated soil structure. Chernozems due to their high mineral contents possess high moisture storage capacities and represent highly fertile black soils of the earth. Chernozems are well known for giving high agricultural yields. In modern Soil Taxonomy, Chernozems - abundant under grassland vegetation - are close to the Order Mollisol. Majority of the Chernozems exhibit a Mollic epipedon.

38. Chesnutt soils – Chesnutt soils are soils of the cool semiarid grassland vegetation that exhibit a light-brown surface horizon gradually grading into a lighter zone and then into a zone of lime accumulation (closer to the surface than Chernozems). Chesnutt soils in comparison to Chernozems have less humus content, less nitrogen and less moisture storage capacity and are thus less fertile than Chernozems. Chesnutt soils are mostly alkaline and come under the Order Mollisol in modern Soil Taxonomy.

39. Chroma of colours - Chroma defines the gradations of the purity of a colour or the apparent degree of departure from neutral grays to white (intensity or brightness). Chroma in essence is the strength of colour within a scale ranging from 0 to 20. In the numerical notation of 2.5 YR 5/6 for a soil colour read from Munsell Soil Colour chart, the digit 6 indicates the chroma of the soil colour.

40. Chromatography - Chromatography is a method of separation of a particular substance from a mixture of many substances, though in most chromatographic tools available now, an analysis component too has been added to the instrumentation to make the tool capable of separation and quantitative assay as well. Differences in properties of the substances like adsorption, electrical charge, polarity, size etc. are exploited for their separation. In chromatography, there are two phases (gas or liquid or solid) viz. a stationary phase, and a mobile phase. The mixture of the substances to be separated is poured, usually in dissolved or suspended state, over the inert nonreactive stationery phase for their adsorption to the surface of the stationary phase. And then, the mobile phase (usually a gas or a liquid) is made to run over the stationary phase. More soluble or more tightly bound substance runs less distance in a given time e.g., one substance requires longer time to travel than another one which is less soluble in the stationary phase or bound more tightly to the stationary phase. Chromatography is therefore, like a running race and the sample components get separated from each other depending upon their solubility in the mobile phase or the degree of binding to the stationary phase. Chromatography can thus be based on partition or adsorption i. e. partition chromatography or adsorption chromatography. Depending on whether partition or adsorption as the principle of separation, and based on the nature of the two phases, chromatography can be of different types such as paper, thin layer, column, gas liquid, gas solid, ion exchange and high-performance liquid chromatography.

41. Coefficient of Linear Extensibility of soils - Coefficient of Linear Extensibility, often referred by its standard abbreviation COLE, refers to the expansiveness of a soil and is indicative of the swelling capacity of a soil. COLE is quantified as

COLE = $[(\frac{L_m - L_d)}{L_d})] \times 100$ where, L_m and L_d represent the lengths of a bar shaped soil when moistened to its plastic limit, and when air dried, respectively.

42. Cohesion - Cohesion refers to the attraction between two similar particles or substances i.e. between two individual water molecules or two individual soil particles. Cohesion involves diverse kinds of bonds.

43. Colloid - The term colloid refers to very small sized particles (10 to 2000 Å) that form a colloidal system made up of a dispersed phase (the colloidal particles) and a dispersion medium. Fog, mist, and milk are colloidal systems (solutions). Clay particles present in soil are classified as colloidal particles, despite however, not satisfying the criteria for sizes.

44. Colluvium - Colluvium is a deposit of soil/soil materials such as stony materials and rock fragments carried down by the action of gravity and deposited at the base of steep slopes such as the bottom of hills. In cool areas, frost action also joins the development of a colluvial deposit.

45. Colorimetry - Colorimetry is an analytical tool based upon the interaction of an incident electromagnetic radiation with matter. In colorimetry, visible radiation in the form of a narrow range of wavelengths (generated by passing the whole light from a source such as a tungsten filament lamp through optical filters) is passed through the sample solution and then the intensity of the radiation absorbed by the solution or the same transmitted through the solution is measured by suitable devices. The extent of absorption or transmission is related to the concentration of the substance in the solution that is to be determined. The interaction of radiation with matter is thus made quantitative and this quantification makes colorimetry a very simple and sensitive analytical tool for the determination of a wide-ranging substances e. g. medicines, pesticides, fertilizers and so on. Colorimetry uses a colorimeter or more correctly speaking a photoelectric colorimeter.

46. Complementary ions - If a system such as a soil does have a definite number of different ions viz. Ca^{2+}, Mg^{2+}, K^+, Na^+ and H^+, and one is interested in only a particular one among them, say H^+, then in the given case, the remaining four ions together viz. Ca^{2+}, Mg^{2+}, K^+, and Na^+ are referred to as complementary ions.

47. Critical Levels of nutrients in plants - Critical Levels (C. L.) of nutrients in plants is the most important approach for the interpretation of the results of plant analysis for diagnosis of nutrient deficiency, and an immediate or a future fertilizer recommendation programme. Critical level of a nutrient in plants is defined as its concentration below which its deficiency symptoms appear and so, yields are decreased. However, in many instances, crop yields decrease much before the visual deficiency symptoms are observed. Critical level of a nutrient in plants currently in use and accepted by the majority concerned is the concentration of the nutrient in plants at which the yield is 90 or 95% of the potential (maximum) yield. Among many aspects in fixing the critical level of a nutrient in crop plants, three most important are species, stage of the crop, and the sampled plant part. As an instance, the critical level of molybdenum in wheat and rice is considered to be 0.3 ppm.

48. Critical Relative Humidity - Critical Relative Humidity (CRH) of a substance refers to the Relative Humidity of the atmosphere at or above which the substance starts absorbing moisture from the atmosphere i.e. exhibits hygroscopicity. CRH of urea and sulphate of potash (SoP) are 72.5 and 96.3, respectively. Therefore, urea unlike SoP, should be stored in dry conditions in rainy season.

49. Darcy's law - According to Darcy's law, the velocity of flow of a fluid, say water (Q), through a porous medium, say soil, is proportional to the force causing the flow and to the hydraulic conductivity (K) of the said medium. Thus, $Q = c\ K\ A\ \frac{h_2 - h_1}{L}$, where c = a dimensionless proportionality constant, A = cross sectional area of the porous medium, h_1 and h_2 = heights above a datum of water and L = length over which the hydraulic drop is taking place. The law is applicable under steady laminar flow conditions when flow is low enough and not turbulent.

50. Deflocculation - Deflocculation is the state of dispersion of a colloidal system wherein the electrically charged colloidal particles repel each other and thus don't tend to come close to each other to coalesce and form floccules. Deflocculation, the reverse of the process of flocculation, is favoured by the large number of water molecules associated with the colloidal micelles and also with the adsorbed cations such as sodium ions.

51. Denitrification - Denitrification, the reverse of the process of nitrification, is a wasteful microbiological process, since considerable amount of fertilizer nitrogen and precious available soil nitrogen is lost in the process, more so in submerged, waterlogged and compact soils. In denitrification, NO_3^- inherently present in the soil or derived from nitrate fertilizers or from NH_4^+ by the process of nitrification is converted to oxides of nitrogen or elemental nitrogen. Four reactions involved in the process are:

(i) $2\ HNO_3 + 2\ H_2 \rightarrow 2\ HNO_2 + 2\ H_2O$; (ii) $2\ HNO_2 + H_2 \rightarrow 2\ NO + 2\ H_2O$

(iii) $2\ NO + H_2 \rightarrow N_2O + H_2O$; (iv) $N_2O + H_2 \rightarrow N_2 + H_2O$

The process need not necessarily continue upto the fourth and last of the four reactions forming molecular nitrogen and can stop at the level of NO. In any case, nitric oxide (NO), nitrous oxide (N_2O), and elemental nitrogen (N_2) escape into the atmosphere, making denitrification a wasteful soil microbiological process causing much loss of applied nitrogenous fertilizers, and of course, farmers' money. The denitrifying microbes (mostly bacteria) are facultative anaerobic microbes e. g. *Thiobacillus denitrificans*. Denitrification is a reduction process and takes place in reduced soils (submerged or compact soils) where oxygen is not available.

52. Desert/Arid soils - Desert/Arid soils are found in areas where the climate is hot and arid, and rainfall is scanty. Wind exerts a prominent role in the formation of desert soils. Soil texture is generally sandy and salt content is high, often to the saline side. Organic matter and nitrogen contents are very low. A *kankar* layer is often found in the lower horizons because of calcium carbonate concretions; the *kankar* layer restricts infiltration of water. Desert soils have very poor crop productivity. In Soil Taxonomy, desert soils belong to the soil Orders Aridisol and Entisol. In India, the desert/arid soils occur in vast areas of the states of Rajasthan, Punjab, Haryana and Gujarat.

53. Desertification - Desertification is defined as a decline in the productivity of the land. Due to desertification, the land can no longer support the same plant growth which it had in the past. The decline in the productivity of the land is often irreversible and permanent on a human time scale. Some of the frequently observed causes that lead to desertification are excessive cultivation, indiscriminate use of chemical fertilizers, overgrazing, excessive withdrawal of groundwater, deforestation, decline in biodiversity, keeping the lands fallow without any vegetative cover, collapse of soil structure, fire, and loss of fertile top soil by erosion. Desertification is a kind of land degradation.

54. Diagenesis - Diagenesis refers to the process of transformation of the deposited rock and mineral materials to form new hard rocks. Diagenesis involves the processes of compaction by the weight of the overlying upper layers and cementation by cementing agents like silica, lime and iron oxides. Such cementation is often called lithification.

55. Diagnostic horizon - Diagnostic horizons are the horizons of soils that have been formed through distinct pedogenic processes. Diagnostic horizons are amply indicative of the soil properties used for identifying and classifying soils. Diagnostic horizons are of use in Soil Taxonomy and are not necessarily consistent with the A, B and C horizons of soils. Based on the depth of location within the soil profiles, diagnostic horizons used in Soil Taxonomy are of two types of viz. diagnostic surface horizons called epipedons and diagnostic subsurface horizons called endopedons.

56. Diffuse Double Layer - A negatively charged colloidal particle like a soil silicate clay behaves like a huge anion and holds a swarm of cations by electrostatic forces. These cations are however, not bound in a single layer around the clay, but are spread in two layers all around the clay. These two layers are collectively called the 'diffuse double layer'. The cations closest to the clay surface are attracted the strongest and form the innermost Stern layer. The rest of the cations are weakly bound and form the outer or diffuse layer, also known as the Gouy layer. Cation concentration is therefore the highest against the clay surface and decreases more and more away from the clay surface. Anion concentration on the other hand, is the least against the clay surface and increases with distance. The boundary of the diffuse double layer is the line where the cation and anion concentration are equal. The thickness of the diffuse double layer is determined by the composition and concentration of the cations. If the external solution contains only monovalent cations with a large hydrated ionic radius (Na^+), the diffuse double layer shall be thicker than when the solution contains divalent cations with a smaller ionic radius (Ca^{2+}). The thickness of the diffuse double layer is determined also by the ion concentration. If the ion concentration is high, the diffuse double layer shall be thin; the reverse is also true.

57. Diffusion - Diffusion is the random movement of the molecules of a gas due to the different partial pressures of the individual gases (say, CO_2 and O_2 of air) in two neighbouring systems such as the atmospheric air and soil air. The difference in partial pressures of CO_2 and O_2 leads to a continuous interchange of the gases between the soil and atmosphere. In this case, O_2 diffuses out from atmospheric air to soil air, while CO_2 diffuses out in the reverse way

58. DRIS - DRIS (Diagnosis and Recommendations Integrated System) is a nutrient ratio concept used for diagnosing nutrient deficiency or toxicity in plants and for fertilizer scheduling. The concept of DRIS was advanced by Beaufils. A number of yield determinate factors related to soil and plants and even others like climate, insects pests, weeds, and diseases are simultaneously considered for a population made of thousands of plants that are later divided into high and low yielding populations for making a diagnosis for improving crop yields in terms of both quality and quantity. The results are furnished in terms of DRIS indices; those in the range -15 (or -10) to +15 (or +10) are considered normal and in balance; those between -25 and -15 are considered a likely deficiency and those > +25 are considered as an indication of the possible nutrient excess. The greater the magnitude of the said index of a nutrient, positive or negative, the more likely the nutrient in question is out of balance in the plant.

59. *Dubba* and *Chalka* soils - The red sandy loam soils widespread in the plains of Telangana and some areas of Andhra Pradesh are locally called *Chalka* soils. Red loamy sand lateritic soils found in patches in the same area are locally referred as *Dubba* soils.

60. Dunes - Dune is a deposit of soil/soil materials such as sand and silt carried out by wind action. Dunes are like ridges of various shapes and common in deserts/arid zones.

61. Earth's Crust - The upper 20 miles of the lithosphere (solid sphere of the earth) is called the earth's crust. The earth's crust is thus a part of the lithosphere. Almost all the elements of the earth are found in the earth's crust. Of all the elements in the earth's crust, eight of them viz. oxygen, silicon, aluminium, iron, calcium, magnesium, sodium and potassium, together constitute as much as 99.6% by mass of the earth's crust upto 16 m. Again, out of these eight elements, two elements viz. oxygen and silicon (both nonmetals) are in very high abundance and these two elements together account for nearly 75% of the mass of the earth's crust. The remaining six of the eight elements viz. aluminium, iron, calcium, magnesium, sodium and potassium, are metals.

62. Edaphology - Edaphology is the study of soils in relation to the growth of higher plants. In edaphology, the soil is thus conceived as a natural habitat for plants. In edaphology, soil properties only of any significance to plant growth and soil productivity are considered.

63. Electromagnetic Spectrum - Light is a form of energy emitted in the form of electromagnetic radiation. The latter moves in the form of waves and do have characteristic wavelengths (distance between a crest and its trough). The wavelength of electromagnetic radiation may be as short as less than one millionth of a nanometer *i. e.* 10^{-15} m (γ rays) and as long as 10^{8} m or more than 62,000 miles (radio waves). The entire range of electromagnetic spectrum so stated is referred to as the electromagnetic spectrum. For common usage, the electromagnetic spectrum is split into distinct regions such as γ rays ($< 10^{-11}$ m), X-rays (10^{-8} m to 10^{-11} m), ultra violet (10^{-7} m to 10^{-8} m), visible (4 x 10^{-7} to 7.6 x 10^{-7} m), infrared (7.6 x 10^{-7} to 10^{-3} m), microwaves (10^{-3} to 10^{-1} m), and radio waves (10^{-1} to 10^{3} m). Analytical tools such as spectrophotometry and flame photometry are based on the interaction of specific regions of the electromagnetic spectrum with matter.

64. Eluviation – Eluviation is an important soil forming process. In eluviation, a washing out process, the constituents of soils like silica, alumina, iron oxides, lime, clay etc. are translocated down by percolating water from the upper to lower soil depths or horizons. Eluviation can be mechanical causing downward translocation of soil particles like silt and clay in a state of suspension or chemical causing downward movement of constituents like silica, silicic acid, alumina, salts, iron oxides, lime etc. in a state of solution

65. Endopedon - The term endopedons as used in Soil Taxonomy are the diagnostic subsurface horizons and include the lower part of the soil where soil materials accumulate, usually due to illuviation. *Argillic, agric, spodic, oxic, kandic, salic, albic, calcic,* and *natric* are some of the important endopedons.

66. Epipedon - The term epipedons as used in Soil Taxonomy are the diagnostic surface horizons and include the upper part of the soil darkened by organic matter. Epipedons are not necessarily the A-horizons of the soils, but the layer of the soil formed at the surface, usually upto 15 to 19 cm. Mollic, Ochric, Histic, and Anthropic are some of the important epipedons.

67. Equilibrium Constant and Law of Mass Action - The ratio of the product of the active masses of the products of a chemical reaction divided by the product of the active masses of the reactants of the reaction, each active mass being raised to a power equal to the number of moles of the reactants and the products, at a point of time when the state of equilibrium of a reversible reaction at a given temperature has been achieved, is a constant known as the Equilibrium Constant or K_{eq} of the given reaction. The statement is also an expression of the Law of Mass Action which states that the rate of a chemical reaction is directly proportional to the active masses of the reactants. For the reaction, a A + b B→ c C + d D, the K_{eq} is

$$K_{eq} = \frac{[C]^c\,[D]^d}{[A]^a\,[B]^b}$$

68. Equivalent weight - The equivalent weight of a substance is a number which states how many parts by weight of it contains or releases or reacts with 1.009 parts by weight of hydrogen or 8.000 parts by weight of oxygen or 35.46 parts by weight of chlorine or an equivalent of the amounts that have been mentioned in the case of substances that do not contain hydrogen or oxygen or chlorine.

69. Equivalent weight of acids - The equivalent weight of an acid is that mass of it which contains 1.009 parts of replaceable hydrogen (H^+). The equivalent weights of acids like HCl, HNO_3, CH_3COOH and H_2SO_4 are 36.5, 63, 60 and 49 respectively, since these many parts of the substances contain 1.009 parts of replaceable H^+.

70. Equivalent weight of bases - The equivalent weight of a base is that mass of it which contains 17.009 parts of ionizable hydroxyl (OH^-) groups or an equivalent of it. The equivalent weights of bases like NaOH, NH_3, $Mg(OH)_2$, and CH_3NH_2 are 40, 17, 29 and 31 respectively, since these many parts of the substances contain 17.009 parts of OH^- or an equivalent of it

71. Essential minerals - Those minerals which are found in rocks in very high amounts so much so that they form the characteristic component minerals of the rocks for their identification and characterization are called essential minerals. Quartz, feldspars, and micas are essential minerals for several rocks.

72. Eutrophication - The high nutrient levels especially nitrogen and phosphorus of water bodies like rivers, streams and lakes encourage excessive growth of algae and other aquatic flora. This boom in algal population causes a depletion of the oxygen level of the water and consequently, the growth of fish and other fauna is drastically affected. Eventually, the flora fauna balance is lost and both fail to survive. The effect is called eutrophication.

73. Exchangeable Sodium Per cent - Exchangeable Sodium Per cent (ESP) of a soil is defined as the per centage of the cation exchange capacity of the soil occupied by the base forming Na^+ ions, both expressed in terms of m.e./100 g soil.

$$ESP = \frac{\text{Exchangeable } [Na^+]}{\text{Cation exchange capacity}} \times 100$$

ESP indicates the relative abundance of the hazardous Na^+ ions and serves as an indicator for sodicity or alkalinity of soils. ESP of alkali and saline alkali soils is > 15.

74. Exfoliation - Exfoliation refers to the process of separation of the outer mass of the rocks in the form of distinct layers consequent to differential expansion and contraction between their inner and outer parts because of temperature variations of the day and night or, between the seasons.

75. Extrusive rocks - Extrusive rocks are igneous rocks formed by the cooling and condensation of magma over the earth's surface. Cooling is thus rapid, and the rocks are cracky and fine grained due to rapid crystallization. Basalt, andesite and rhyolite are important extrusive rocks. Extrusive rocks are also known as volcanic rocks.

76. Fick's law of Diffusion - According to Fick's Law of Diffusion, the molar flux of a gas across a plane due to diffusion (N, the diffusive flux) is proportional to the concentration gradient of the gas ($\frac{d_c}{d_x}$). Thus,

$N = - D\frac{dc}{dx}$ where, D is the diffusion coefficient. Since the gradient is towards the direction of lower concentration, $\frac{dc}{dx}$ is assigned a negative sign. The values of D for elemental oxygen (O_2) in air and water are 2.3 x 10^{-1} and 2.6 x 10^{-5} cm^2/s; the same for CO_2 are 1.8 x 10^{-1} and 2.0 x 10^{-5} cm^2/s. Therefore, (i) for the same concentration gradient, the diffusive flux of O_2 is nearly 10^4 times faster through air-filled pores than through water-filled pores, and (ii) O_2 diffuses slightly faster than CO_2 under identical conditions.

77. Field capacity - Field capacity is the amount of moisture retained by the soil after the excess gravitational water has practically drained off. At field capacity, water moves out from macro pores to be occupied by soil air, while the micro pores are still filled with water to be available to plants. The soil moisture tension at field capacity corresponds to a pF of 2.53 or $\frac{1}{3}$ bar.

78. Flame photometry – Flame photometry is an analytical method for the determination of some metallic elements. In flame photometry, the substance to be determined is thermally dissociated in a flame to its constituent atoms. Some of these atoms, provided they are of alkali and alkaline earth metals, are further excited to their higher energy levels by the flame energy. Later, when these excited atoms return back to the ground state (the normal and most stable atomic orbital configuration of an atom), the absorbed energy is emitted in the form of electromagnetic radiation, the wavelength of which is a characteristic of an element. This emitted particular wavelength radiation (589 and 766.5 for sodium and potassium, respectively) is quantified and correlated to the concentration of the radiation emitting element present in the substance. Flame photometry is emission spectrophotometry and is particularly useful for the determination of sodium, potassium, calcium and magnesium.

79. Flocculation - Flocculation refers to the process of aggregation or clumping together of the small solid particles of the dispersed phase of a colloidal system in the form of floccules (aggregates). In soils, flocculation is the first step towards the formation of stable soil aggregates or granules required for the formation of most favourable granular or crumby soil structure for plant growth.

80. Fourier's law - Fourier's law states that the heat flow by conduction in a homogenous body, say soil, is in the direction of and proportional to the temperature gradient. Thus,

q_h = - k ∇T, where q_h, k and ∇T denote the heat flux, thermal conductivity and spatial gradient of temperature T, respectively. In one dimensional form, the law is written as

$q_h = - k \times \frac{dT}{dx}$, where $\frac{dT}{dx}$ is the temperature gradient in any arbitrary direction designated as x.

81. Gapon equation – Gapon equation has its origin in the cation exchange behaviour of soils. Consider the following exchange reaction of NH_4^+ in soil solution with Ca^{2+} on the soil exchange complex.

(NH_4^+) + (Ca-exchanger) ⇌ ½ (Ca^{2+}) + (NH_4-exchanger), where (NH_4^+) and (Ca^{2+}) represent molar activities of the ions in soil solution, and (Ca-exchanger) and (NH_4-exchanger) represent the activities of the ions on the soil exchange complex. Applying the law of mass action, the equilibrium constant (K_{eq}) of the exchange reaction is

$$K_{eq} = KNH_4.Ca = \frac{(NH_4 - \text{exchanger})\sqrt{(Ca^{2+})}}{(Ca - \text{exchanger})\,(NH_4^+)}$$

If now, the ion activity ratio on the exchange complexes is assumed equal to the equivalent fractions, and concentration terms (represented by square brackets) are substituted for the activities of the ions in solution, the equation can be rearranged as follows.

$$\frac{NH_4\text{–exchanger}}{Ca\text{–exchanger}} = KNH_4.Ca * \frac{[NH_4^+]}{\sqrt{[Ca^{2+}]}}$$

The above equation is an empirical expression of Gapon equation in which $KNH_4.Ca$ is called the Gapon coefficient. Ca^{2+} and Mg^{2+} have sufficiently close adsorption affinities relative to monovalent cations like NH_4^+, Na^+ and K^+. Therefore,

$$\frac{Na\text{–exchanger}}{Ca.Mg\text{–exchanger}} = KNa.Ca.Mg * \frac{[Na^+]}{\sqrt{[Ca^{2+} + Mg^{2+}]}} \quad \text{and}$$

$$\frac{K\text{–exchanger}}{Ca.Mg\text{–exchanger}} = KK.Ca.Mg * \frac{[K^+]}{\sqrt{[Ca^{2+} + Mg^{2+}]}}$$

The equation is of much practical utility in determining the availability of plant nutrients in a heterogenous exchange system such as a soil and also in predicting the likely sodium hazard of irrigation water. The expression for Sodium Adsorption Ratio (SAR) has its origin in the equation.

82. Gleization - Gleization is a process of soil formation resulting in the development of a 'glei' (glei refers to blue, gray or green clay) in the lower part of the soil profile due to poor drainage and waterlogged conditions. Such soils are called 'hydromorphic' soils. Local conditions or microclimate rather than the climate leading to poor drainage are important for the formation of such soils. Under anaerobic conditions, soluble ferric iron (Fe^{3+}) compounds are reduced to soluble ferrous iron (Fe^{2+}) compounds. The reduction is primarily biological and requires both organic matter and microorganisms capable of respiring anaerobically. Reduced Fe exists as Fe^{2+}-organic complexes in solution or as a precipitate of ferrous hydroxide which is responsible for the formation of a typical bluish to grayish soil horizon. Due to seasonal fluctuations of water table, the gley shows distinct mottling of yellow and rusty brown colours caused by alternate oxidation and reduction phenomena.

83. Gypsum Requirement of alkali soils -The amount of gypsum required for lowering the sodium content of alkali or sodic soils to a desired level in relation to those of calcium and magnesium contents is the gypsum requirement (GR) of such soils (the reclamation of alkali soils). Gypsum so added lowers the exchangeable sodium per centage (ESP) and sodium adsorption ratio (SAR) of such soils to a level where plant growth is no longer interfered. Gypsum Requirement of alkali soils is expressed in terms of m. e. of gypsum/100 g soil or in terms of tons/ha or acre for practical field use. Schoonover devised a simple laboratory method for the determination of GR of alkali soils.

84. Hardy Schulze Rule – Hardy Schulze Rule states that greater the valency of the oppositely charged ion of the electrolyte being added to a colloidal solution, faster is the coagulation of the colloidal system. Thus, the coagulation of a negative sol is directly related to the valency of the cation of the electrolyte added and *vice versa.* In the coagulation of a negative sol, the flocculating power of the cations follows the order:

$Al^{3+} > Ba^{2+} > Na^{+}$

In a similar way, in the coagulation of a positive sol, the flocculating power of the anions follows the order:

$[Fe(CN)_6]^{4-} > PO_4^{3-} > SO_4^{2-} > Cl^{-}$

85. Heavy metals – There is no total consensus on the definition of heavy metals. However, a generally accepted definition of a heavy metal is a metal element that is heavy, precisely five times heavier than water. Metals such as mercury, cadmium, nickel, cobalt, zinc, chromium, thallium, lead, iron, copper, silver, and gallium are few in the list of heavy metals. Heavy metals are toxic to animals. With the assumption that heaviness and toxicity of elements are interrelated, some metalloids such as arsenic are also treated as heavy metals. In recent years, heavy metals have been branded as potential environmental pollutants. Soil, water sources like rivers, lakes and ponds, groundwater and air are getting increasingly contaminated by heavy metals. Heavy metals have been reported to cause poisoning in man and animals by disruption of the gastrointestinal systems, inhibition of mitochondrial activities and respiration, inhibition of protein biosynthesis machinery, damage to DNA metabolism, mutagenesis, carcinogenicity and many more. Some of the sources of heavy metals in the environment are mine tailings, industrial discharges, burning of fossil fuels, automobile exhausts, and urban rubbish. e-Wastes like disposed computer chips and cell phones are also rich sources of heavy metals.

86. Hidden hunger and luxury consumption of plant nutrients – Hidden hunger of plant nutrients refers to a situation where plants require a nutrient for their required metabolic functioning and growth as evidenced by the level of the nutrient in the plants below its critical concentration and yet no deficiency symptoms are expressed to indicate the nutrient deficient in the plants. As no deficiency symptoms of the nutrient occur, the latter goes unnoticed and is not applied. Hidden hunger thus causes serious reduction in crop yields. Fertilizer scheduling based on the results of plant analysis can help solve the problem of hidden hunger. By contrast, luxury consumption of a nutrient refers to a situation where a nutrient has reached the optimum and sufficiency level for its functioning and maximum yield in the plant and yet the plant continues absorbing the nutrient from the soil with no additional yield. However, for luxury consumption of a nutrient to take place, the available nutrient content of the soil must be in excess of its requirement. Luxury consumption of potassium is often noticed in crop plants, especially forage crops. Fertilizer scheduling must be based on soil test and plant analysis results to prevent the wasteful process.

87. Hue of colours - Hue of colours refers to the dominant spectral colour or quality which distinguishes red from yellow and so on. Hue is assigned a numerical value e.g. the red soils have hues of 7.5 YR (or, redder) in upper topographic positions.

88. Humus - Humus is defined as a complex and rather resistant mixture of brown or dark brown amorphous and colloidal organic substances modified from the original tissues or synthesized by various soil organisms. Most of the colloidal properties of soil organic matter are due to the presence of humus. Humus is non-crystalline, amorphous, and does posses high surface area, negative charge, cation exchange capacity and many functional groups of which carboxyl (-COOH), phenolic OH and enolic OH groups are the most significant.

89. Hydration - Hydration refers to the absorption of water by a substance in such manner that the absorbed water becomes an integral part of the molecular makeup of the substance. Thus, $CaSO_4$ on hydration forms gypsum ($CaSO_4.2\ H_2O$). Hydration is an important reaction in the chemical weathering of rocks and minerals leading to the formation of soils.

90. Hydraulic conductivity - Hydraulic conductivity of a porous medium such as a soil refers to the ease with which the soil pores permit water movement (downward, horizontal, and even upward) and is described as $V = K f$ where, V, f and K denote the volume of water moved per unit time in the soil's saturated state, water moving force and hydraulic conductivity of the soil, respectively. Hydraulic conductivity of coarse textured soils is higher than that of fine textured soils.

91. Hydrolysis - Hydrolysis refers to the decomposition of a substance by the addition of water. Hydrolysis is like a double decomposition reaction, where the H and OH from H_2O are added to a substance to split it into two molecules of two different compounds. Hydrolysis is considered as the most important chemical reaction in the chemical weathering of rocks and minerals leading to the formation of soils.

92. Hygroscopic coefficient – Hygroscopic coefficient represents the soil moisture content held by the soil in vapour phase. Hygroscopic coefficient depends on the amount and type of clay, the nature of the exchangeable cations, and the presence of free electrolytes. Soil water at this point is not normally available to plants. The soil moisture tension is -31 bar (approximately equal to a pF of 4.54).

93. Igneous rocks - Rocks formed by the cooling down of lava or molten magma produced by volcanic eruptions are called igneous rocks. Igneous rocks are massive in size. They are also the most abundant rocks and account for 95% of the total rock mass of the earth. Granite, basalt, and gabbro are three examples of igneous rocks.

94. Illuviation - Illuviation is an important soil forming process. Illuviation is the process of accumulation of soil materials washed away from the upper soil layers at lower depths of the soils. The washed out soil materials include silica, sesquioxides, lime, fine silt, clay etc.

95. Immobilization – Immobilization refers to the biological process – usually carried out by heterotrophic micro flora and micro fauna - in which a plant nutrient present in simple and plant usable inorganic forms such as ammonium, nitrate, sulphate etc. is converted to complex and plant unusable complex organic forms such as proteins, polysaccharides etc. by the process of reduction. Immobilization is the reverse of the process of mineralization.

96. Intrusive rocks – Intrusive rocks are igneous rocks formed by the cooling and condensation of magma at considerable depth below the earth's surface. Cooling is thus slow, and the rocks are hard, massive and coarse grained. Granite, gabbro and syenite are important intrusive rocks. Intrusive rocks are also known as plutonic rocks.

97. Ion Exchange Capacity – Ion exchange capacity of a charge (positive or negative or both) active substance (such as soil clays) is defined as a measure of its ability to permit displacement of the ions held to it by electrostatic forces by the oppositely charged ions present in the surrounding medium (soil solution in case of soil clays). The displacement of ions between the two phases must however, be on charge equivalence basis (one Ca^{2+} for two H^+ ions or one PO_4^{3-} for three Cl^- ions).

98. Ionic Product of water – The equilibrium constant (K_{eq}) for the dissociation reaction of water is

$$K_{eq} = \frac{[H^+]\,[OH^-]}{[H_2O]}$$

Water is an extremely weak electrolyte and so its molar concentration ($[H_2O]$) is practically $\frac{1000}{18} = 55.5$ M. The same of both H^+ ($[H^+]$) and OH^- ($[OH^-]$) is 10^{-7} M (at 25^0 C)

$$\therefore K_{eq} = \frac{[H^+]\,[OH^-]}{[H_2O]} = \frac{10^{-7} \times 10^{-7}}{55.5} = \frac{10^{-14}}{55.5}$$

$$\text{or, } K_{eq} \times 55.5 = 10^{-14}$$

In the above expression, the left side term (55.5 x K_{eq}) is replaced by a conglomerate constant K_w, called the Ionic Product of Water. The value of K_w at 25^0 C is therefore 10^{-14}. Obviously, in neutral solution, concentrations of H^+ and OH^- are equal at 10^{-7}; in acid solution, the concentration of H^+ is higher and the concentration of OH^- to that extent is lower. The reverse is also true for an alkali solution. The K_w forms the basis for pH scale as a means of expressing the acidity or the alkalinity of a solution.

99. Jarusov's rule – Jarusov's rule, many referring it as the Principle of Complementary Ions, states that the exchangeability/replaceability of an ion present over a charge active substance (like a soil colloid) by another one present in the surrounding solution (soil solution) is not absolute or constant; rather, it is a function of its amount in relation to the amount and kind of all the other ions present in the environment.

100. Lambert Beer's law – Whenever a radiation, either monochromatic or heterogeneous, falls upon a transparent medium or passes through a solution, a part of the incident radiation is absorbed by the medium or the solution, while the rest is either reflected or transmitted. The intensity of the incident radiation is thus the sum of the intensities of the absorbed, reflected and transmitted radiation. According to Lambert Beer's law, when monochromatic light passes through a transparent medium or a solution, the intensity of the transmitted light decreases with an increase in the thickness of the medium or the concentration of the solution. Further, the intensity of the transmitted light decreases exponentially, while the thickness of the medium or the concentration of the solution increases arithmetically. Colorimetry and spectrophotometry are analytical tools based upon the principles of Lambert Beer's law.

101. Land Capability Classification – Land Capability Classification (LCC) is a system of grouping soils (land) into various classes based on inherent limitations imposed on their sustained use by soil attributes, topography, drainage and climate. 'Use land according to its capability and treat it as per its need' is the guiding principle of LCC. In LCC, lands are grouped in eight classes viz. Class I to Class VIII. CLASS I – shown in green colour on maps. Very good cultivability. CLASS II - shown in yellow colour on maps. Good cultivability. Slight limitation due to factors like depth, salinity, texture, drainage and/or erosion. CLASS III - shown in brown or pink colour on maps. Moderately good cultivability. Moderate degrees of limitations due to erosion, depth, salinity, texture, drainage and/or vertic properties. Not suitable for growing vegetable crops. Intensive and careful management practices are required. CLASS IV - shown in blue colour on maps. Fairly good cultivability. Limitations are still higher. Only selective crops can be grown and intensive soil conservation management practices are required. CLASS V, VI, VII and VIII - shown in dark green, orange, red and purple colour respectively, and are not suitable for growing crops. Land under these Classes can be used for grazing, forestry, wildlife and watershed. Each Class of LCC is subdivided and subsubdivided into a number of capability subclasses and capability units.

102. Land Degradation – Land Degradation is defined as the decline in quality of the land caused by a reduction in soil fertility, biological productivity, biodiversity and overall soil health. Land degradation is caused primarily by callous short-term benefits, and thoughtless and irrational human activities which include among others excessive cultivation, indiscriminate use of chemical fertilizers, overgrazing, excessive withdrawal of groundwater, deforestation, decline in biodiversity of the area, keeping the lands fallow without any vegetative cover, collapse of soil structure, loss of fertile top soil by erosion etc. The terms land degradation and soil degradation are used as synonyms.

103. Landfills – Landfills refer to the practice of dumping diverse kinds of rubbish, refuses and waste materials such as household wastes, municipal wastes, industrial discharges and effluents, hospital and health facility wastes etc. in low lying areas, at times by burying underneath the soil, in an otherwise large flat area so that in course of time, the low lying areas come up as upland areas, and be used for various purposes such as parks and recreational areas, shopping complexes, and office buildings. Landfills should however, be very strictly monitored by environmental protection agencies for reasons such as the release of toxic gases, and contamination of the land area and ground water by obnoxious substances such as persistent pesticides, heavy metals, and radioactive substances.

104. Laterites and lateritic soils - Laterites and lateritic soils are highly weathered soils with laterization as the dominant soil forming process. In laterization as is known, the process taking place is eluviation of silica and enrichment of the upper soil layers with the oxides of Fe and Al under conditions of high rainfall and alkaline pH due to basic parent material. These soils are thus rich in sesquioxides, poor in bases, and become hard when exposed to drying after wetting. The material when wet can be cut into definite shapes and be used as bricks, and hence the name 'laterite' (Gr. *later* means bricks). Because of intense weathering, kaolinite is the dominant clay mineral of these soils. The soils are thus low in CEC and percent base saturation. Laterites and lateritic soils are moderately fertile. In Soil Taxonomy, the laterites and lateritic soils belong to the soil Orders Oxisol, Ultisol and Inceptisol. Rice, sugarcane, ragi, coconut, cashew nut, coffee, tea, tapioca and rubber are grown extensively in such soils. In India, laterites and lateritic soils are found on the hill tops and plateau landforms of Kerala, Tamil Nadu, Telangana, and Odisha. In some parts of Telangana, the laterites and lateritic soils are locally referred to as *Dubba* and *Chalka* soils.

105. Laterization - Laterization is an important soil forming process. In laterization (Latin '*later*' meaning bricks) soil formation process, silica is leached down from the upper layers to be deposited in the lower layers under conditions of warm humid climate of the tropics and subtropics, basic parent material, alkaline pH of the environment and rain forest vegetation. Obviously, in such soils, the upper eluvial horizon is rich in sesquioxides and poor in silica; the reverse is the case with the lower illuvial horizon.

106. Leaching Requirement for reclamation of saline soils - The Leaching Requirement (LR) for reclamation of saline soils is the ratio of the equivalent depth (D) of the drainage water to the depth of the irrigation water (D_{dw}/D_{iw}) and may be expressed as a fraction or as percent. The ratio is equal to the inverse ratio of the corresponding electrical conductivities (EC_{iw}/EC_{dw}) Thus,

$$LR = \frac{D_{dw}}{D_{iw}} = \frac{EC_{iw}}{EC_{dw}}$$

Obviously, for the estimation of LR, the salinity level of the irrigation water (EC_{iw}) and that of the drainage water (EC_{dw}) must be known. For a field crop which can tolerate an EC_{dw} of 8 dS/m, LR with irrigation water having electrical conductivities of 1, 2, 3 and 4 dS/m is 12.5, 25, 37.5 and 50 per cent, respectively. These are however, the maximum values, since rainfall, removal of the salts by the crop, and precipitation of the salts such as calcium carbonate or gypsum in the soil, and few more such factors have their own contributions, duly lowering the theoretically calculated value of LR.

107. Lime Potential – Several factors such as the suspension effect, soil:water ratio, electrolytes concentration etc. lead to variable results in soil pH measurements. To overcome the effect, pH measurements are done in 0.01 M $CaCl_2$ solution. In this context, Schofield and Taylor (1955) suggested the use of ion activity ratios for the determination of soil acidity and base saturation instead of using single ion measurement. Assuming a soil exchange complex to remain saturated with only H^+ and Ca^{2+} ions, we have at equilibrium according to Schofield's Ratio Law,

$\frac{H^+}{\sqrt{Ca^{2+}}}$ = constant, which on taking negative logarithm (p = -log) changes as

$$-\log \frac{H^+}{\sqrt{Ca^{2+}}} = \text{constant}$$

or, $-\log H + \frac{1}{2}\log Ca^{2+}$ = constant

or, pH - ($-\frac{1}{2}$ log Ca^{2+}) = constant

or, pH $-\frac{1}{2}$ p Ca^{2+} = constant (since, $-\frac{1}{2}$ log Ca^{2+}= $\frac{1}{2}$ p Ca^{2+})

This expression is called the lime potential. Lime potential characterizes the composition of the exchange complex relative to its saturation by H^+ and Ca^{2+} ions. Many methods for the determination of lime requirement of acid soils take into count the lime potential of soils. The equation states that a change in soil pH is dependent on Ca^{2+} concentration expressed as the calcium potential i.e. pH = $\frac{1}{2}$ p Ca^{2+} + constant.

108. Lime Requirement of acid soils - Lime Requirement (LR) is defined as the amount of a liming material that is to be added to an acid soil for its reclamation e. g. for increasing the existing soil pH to a desired level for optimum plant growth and crop yield. It is usually expressed in terms of tons/acre or ha. Lime requirement of acid soils is directly related to their cation exchange capacity which in turn depends on the clay content, nature of the clays and organic matter content of the soils.

109. Lithosphere – Lithosphere is one of the three spheres of the earth. Of the three spheres, the solid sphere comprised by the continents, ocean basins, plains, plateaus, mountains, valleys, sand dunes, lava flows, and fault scraps and covered by the atmosphere and/or hydrosphere is referred as the lithosphere. Rocks, minerals, and soils are parts of the lithosphere.

110. Loess – Loess is a deposit of soil/soil materials such as fine silt and clay transported and piled up by the action of wind. Loess is usually weakly compacted and grey or yellowish in colour.

111. Lyotropic series – Lyotropic series suggested by Franz Hofmeister, also known as Hofmeister series, is a series of ions arranged in an order (ascending or descending) of the relative ease of the replaceability of the ions adsorbed over a charged surface. For cations, the order of the lyotropic series is explainable, if the cations at the colloid surface include a layer of specifically adsorbed or partially dehydrated cations, the so called Stern layer. The order of replaceability of the cations adsorbed over the surface of montmorillonite has been formulated as follows (there are however, reports with a marginally different order)

$Li^+ \approx Na^+ > K^+ \approx NH_4^+ > Rb^+ > Cs^+ \approx Mg^{2+} > Ca^{2+} > Sr^{2+} \approx Ba^{2+} > La^{3+} \approx H^+$ $(Al^{3+}) > Th^{4+}$

As in the case of the cations, for the anions too, there is a lyotropic series applicable for positive sols. Thus,

$CO_3^{2-} > SO_4^{2-} > H_2PO_4^- > Cl^- > NO_3^- > ClO_4^- > SCN^-$

In either series, the ions on the left side are well hydrated and tend to stabilize a sol (kosmotropic ions), while those on the other side are poorly hydrated and tend to flocculate a sol (chaotropic ions).

112. Magma - The hot complex silicate melt of the lithophilic elements (containing water vapour and temperature of 1400^0 C or more) underneath the earth's surface is called the magma. Cooling of the magma over the surface of the earth or underneath forms igneous rocks.

113. Mass flow - Mass flow refers to the movement of a fluid from one system to another down a concentration or pressure gradient. Mass flow of gases into and out of two systems, say soil air and atmospheric air, occurs due to the difference in total pressure (pressure gradient) between the soil air and atmosphere. This pressure gradient caused by expansion and contraction of the soil gases due to changes in temperature and pressure, and air replacement by drainage etc. causes movement of the entire mass of air from a zone of high pressure to a zone of low pressure. Flow of gases due to mass flow may also occur from one narrow pocket of a soil to another narrow pocket due to similar reasons.

114. Melanization – Melanization is a process of darkening of the soils due to the addition of organic matter. Darkening of the soils is predominantly a product of the activities of the grasses and earthworms. The process of melanization is prominent in fertile dark crumb-textured grassland soils of the soil order Mollisol.

115. Metamorphic rocks - Those rocks that have been formed by the action of heat and pressure on igneous and sedimentary rocks resulting in metamorphosis (visual and inner changes) of the rocks – so much so that the original rocks are altogether altered – forming new kinds of rocks are called metamorphic rocks. Mineralogical composition of the metamorphic rocks depends upon the composition of the original rocks and the kind of metamorphism. Marble, slate and diamond are three of the many examples of metamorphic rocks.

116. Mineral - A mineral is a naturally occurring substance that has more or less definite physical and chemical properties such as colour, lustre, cleavage, fracture, crystal structure, crystal geometry, hardness, refractive index, specific heat, specific gravity, optical activity, magnetic behaviour etc. (all physical properties), and chemical composition, reactivity, chemical stability etc. (all chemical properties). Gypsum, hematite, quartz, corundum, topaz, calcite, and epsomite are few important minerals.

117. Mineralization – Mineralization refers to the biological process - usually carried out by heterotrophic micro flora and micro fauna - in which a plant nutrient present in complex organic and plant unusable forms such as proteins and lignin is converted to simple inorganic and plant usable forms such as ammonium, nitrate, sulphate, phosphate etc. Mineralization is an oxidation process and is the reverse of the process of immobilization.

118. Mohs scale – Mohs scale is a 0 to 10 point scale devised by Friedrich Mohs for measuring the hardness of minerals. Mohs scale is indicative of a relative measure of the mineral's resistance to scratching. The said resistance to scratching is quantified by scratching the mineral against another substance of known hardness in the scale. Ten minerals with Mohs scale values of 1 to 10 based on the relative ease or difficulty with which one can be scratched by another are talc- 1, gypsum- 2, calcite-3, fluorite-4, apatite-5, orthoclase-6, quartz-7, topaz-8, corundum-9 and diamond-10. It is worth mentioning in the context that some of the objects used for scratching are fingernail (2.5), copper penny (3.5), knife/glass plate (5.5), steel nail (6.5) and masonry drill belt (9.5).

119. Moist bolus – The moist soil mass used for ascertaining an approximate qualitative estimate of the texture of the soil by making a soil ribbon by rolling the moist soil mass over the palm with the help of fore fingers and then judging its strength for its texture is called the moist bolus.

120. Mole – Mole is defined mole as the amount of a substance which contains as many elementary units like the atoms, molecules, ions, radicals, or such particles as there are atoms in 0.012 kilogram of Carbon-12 (^{12}C) which equals to 6.02 x 10^{23} atoms of C^{12}. One mole of H^+, Ca, NH_3, CO_3^{2-} and Na_2CO_3 consists of 1, 40, 17, 60 and 106 g of H^+, Ca, NH_3, CO_3^{2-} and Na_2CO_3, respectively.

121. Mole fraction - Mole fraction is defined as the ratio of the number of moles of a substance to the total number of moles of all the substances present in the homogenous system such as a solution.

122. Mor - Mor is a type of humus rich surface horizon developed under heavy forest vegetation (usually coniferous) acid litter, where the activity of acidophilus fungi is the maximum and bacterial activity is at its minimum. Mor is composed of several layers of organic matter in different degrees of decomposition.

123. Mulching - Mulching is the application or creation of a soil cover by the application of specific materials that constitutes a barrier to the transfer of heat and loss of water vapour from the soil. Paper, straw, hay, sugarcane trash, water hyacinth removed from ponds and tanks, saw dust, farm yard manure and plastic sheet are some of the commonly used such materials called mulches. Mulches conserve soil moisture and buffer the extremes of soil temperature.

124. Mull – Mull is designated as a forest (usually coniferous) surface soil horizon intimately mixed with mineral matter and is slightly acidic. Mull is best developed under base rich litter, where among the soil microbes, bacteria and fungi play a significant role.

125. Mycorrhizae – The symbiotic association established by certain fungi with thc cells of the roots of certain higher plants is called mycorrhizae. The fungi secure sugars (C) and energy requirements from the root cells of the plants, and in return help extracting phosphorus and some micronutrients from the soils for the benefit of the host plant, thus establishing an active and mutually beneficial symbiotic association. Mycorrhizal association is found in many agronomic crops.

126. Nano fertilizer – Nano fertilizers are fertilizer molecules, engineered physically or chemically or in both ways, containing one or more plant nutrients. Nano fertilizers are thus tailor-made fertilizer molecules secured by techniques that include among others physical and chemical modification, encapsulation, coating, and reduction in particle size upto nano level of the existing nutrient molecules. The specially designed nutrient molecules called nano fertilizers ensure controlled and sustained release over an extended period of time and slow diffusion of them into the soil system matching the plants' needs. There is practically zero loss of the fertilizers by leaching, runoff, volatilization and nonspecific use, thus resulting in exceedingly high recovery of the applied plant nutrients and reducing environmental hazards associated with the application of chemical fertilizers. Kalol unit of Indian Farmers Fertilizer Cooperative Limited (IFFCO) was the first to launch the commercial production of nano fertilizer called NANO UREA Liquid (India's first nano fertilizer, 4% N, 500 ml bottle). Particle size of nano urea is 20 to 50 nm in comparison to that of 1 mm of prilled urea and thus there are 55000 nitrogen particles over a 1 mm urea prill. It has been claimed that a 500 ml bottle of nano urea can replace one bag of conventional urea. After nano urea, nano DAP is also on the anvil in Government of India's ambitious plan of running with nano technology in the field of agriculture. There is however, some controversy too with claims and counterclaims for the performance of nano fertilizers.

127. Nitrification - Mineralization of organic nitrogen produces nitrogen in ammonical form (NH_4^+). The latter may be derived also directly from ammonical fertilizers or from the hydrolysis of urea by the enzyme urease as shown below.

$$CO(NH_2)_2 + H_2O \rightarrow 2\ NH_3 + CO_2$$

This NH_3 is then converted to nitrate (NO_3^-) by the process of nitrification. The NO_3^- is then absorbed by plants. Nitrification is a beneficial microbiological process carried out in two steps. In the first step, NH_4^+ is oxidized to NO_2^- (nitrite) by *Nitrosomonas* group of bacteria. In the second step, NO_2^- is oxidized to NO_3^- by *Nitrobacter* group of bacteria. Thus,

$$1^{st}\ \text{step: } 2\ NH_3 + 3\ O_2 \rightarrow 2\ HNO_2 + 2\ H_2O$$

$$2^{nd}\ \text{step: } 2\ HNO_2 + O_2 \rightarrow 2\ HNO_3$$

Specific enzymes are involved in both the steps. Both *Nitrosomonas* and *Nitrobacter* are chemoautotrophic soil bacteria. Nitrification is an oxidation process and thus takes place in aerated soils where oxygen is available. Nitrate is however, highly mobile and a significant pool of it is lost by leaching much before the plants absorb it. To prevent this loss, chemicals called nitrification inhibitors are used. They are inhibitory only to *Nitrosomonas* bacteria. Important among them are N-Serve, AM and nitrapyrin.

128. Origin of negative charge in humus – Humus is the soil's organic colloid and is negatively charged. The negative charge in humus arises from the dissociation of the functional groups of the organic compounds present in humus. Important among these functional groups are carboxyl (-COOH), phenolic OH (-C_6H_4OH), and enolic OH groups. At low pH, these functional groups remain undissociated with no charge. As pH increases (more of OH^- ions being added), first the –COOH groups, then the phenolic OH groups, and then the enolic OH groups dissociate forming –COO^-, -$C_6H_4O^-$ and enolic -O^- anions, respectively, and H^+ ions which unite with OH^- ions forming H_2O molecules. Humus thus develops negative charge from the dissociation of these functional groups. This kind of negative charge is called pH dependent negative charge or variable negative charge. Humus doesn't have negative charge derived from isomorphous substitution (pH independent or permanent charge). Further, for electroneutrality, this negative charge remains neutralized by the positively charged cations drawn from the surrounding medium. These cations are exchangeable on charge basis by other cations present in the same surrounding. The process is the cation exchange process and the capacity of the humus it can lend its cations for exchange with other cations in the surrounding solution is the cation exchange capacity of humus.

129. Origin of negative charge in silicate clays - There are two principal ways for the source of negative charge in silicate clays viz. derived from the exposed or broken edges, and isomorphous substitution. The silicate clays develop a negative charge from the oxygen ions present along the edge of the crystals. For these oxygen ions, only one negative charge is satisfied by crystal ions, while the other one remains unsatisfied. There are thousands of such oxygen ions all along the edges of the crystals. At very low pH when H^+ ions are abundant, these negative charges remain neutralized by H^+ ions forming OH groups. As pH increases, these OH groups dissociate forming the $-O^-$ ions (and H^+ ions), much like a weak acid e. g. CH_3COOH. Such dissociation thus creates negative charge in the crystals. For obvious reason, the magnitude of such negative charge increases with increase in pH and hence such negative charge is called pH dependent negative charge or variable negative charge or temporary negative echarge. Obviously, the CEC of inorganic soil colloids increases with increase in pH. For 2:1 type clay minerals, edge surface is small and hence pH dependent charge is not appreciable in comparison to the small particle size. In 1:1 clay minerals like kaolinite, the minerals tend to stack along the c-Axis due to which edge surface is high and almost the entire negative charge (and CEC) is derived from pH dependent charge. The other source of negative charge in silicate clays is the isomorphous substitution which is the substitution of one ion by another one similar in ion geometry. If the two ions are of different valency, a charge is created in the crystals. Thus, if Si is substituted by Al (ionic radii of Si^{4+} and Al^{3+} are 0.042 and 0.051 nm – largely similar and hence one can fit snugly in the other's place), one unit of negative charge is created in the crystals (because tetravalent Si^{4+} is being replaced by trivalent Al^{3+}). Similar is the case of trivalent Al^{3+} by divalent Mg^{2+} (ionic radii of Al^{3+} and Mg^{2+} are 0.051 and 0.064 nm – largely similar). The reverse of the case is also possible creating positive charge in the crystals. Such substitution is possible to the extent of 15% without disturbing the structure of the crystal lattice. Development of such kind of charge is in no way related to pH. Such kind of charge is therefore referred to as pH independent charge or permanent charge. The total charge of silicate clays is the sum of the pH dependent charge (temporary or variable charge) and pH independent charge (permanent charge).

130. Origin of positive charge in silicate clays – Silicate clays do have pockets of positive charges also. There are two sources of such positive charges viz. protonation and isomorphous substitution. Silicate clays develop positive charges due to protonation of the exposed OH groups under moderate to extremely acidic pH (more of H^+ ions). The OH groups which at high pH form negative charges develop positive charges at low pH due to protonation. Protonation is a process in which protons (H^+) are attached to the surface of the OH groups forming OH^+ ions. The magnitude of such positive charge increases with decrease in soil pH. Such positive charge depends upon pH and hence is thus called pH dependent positive charge or variable positive charge. In isomorphous substitution, a cation of a higher valency substitutes a cation of lower valency. The result is the silicate clay crystal with a positive charge. For instance, in a trioctahedral sheet, there are three Mg^{2+} ions for two oxygen and two hydroxyl ions. The sheet thus has no positive charge. If, however, one Mg^{2+} ion is replaced by an Al^{3+} ion, the sheet develops one unit of positive charge. This kind of charge is not affected by pH and is called pH independent charge or permanent charge.

131. Oxidation - The term oxidation in its narrowest sense refers to a chemical reaction in which oxygen is added to or hydrogen is removed from a substance. In more refined terms, in oxidation, there is a loss of one or more electrons suffered by an atom or ion or a molecule. The atom or ion or molecule in the case is said to have undergone oxidation i.e. be oxidized. Conversion of H_2O to H_2O_2 by the addition of oxygen, Fe^{2+} (ferrous iron) to Fe^{3+} (ferric iron) by loss of an electron and I^- (iodide) to $\frac{1}{2}$ I_2 (elemental iodine) by loss of an electron are three simple examples of the process of oxidation. Oxidation is the reverse of the process of reduction. In reality, oxidation and reduction take place simultaneously implying that while one reactant undergoes oxidation, another one undergoes reduction. The reactant that undergoes oxidation forms the reducing agent, while the reactant that undergoes reduction forms the oxidizing agent. Further, there may be other reactants too in an oxidation reduction reaction which facilitate the processes of oxidation and reduction to proceed. Oxidation is an important process in the chemical weathering of rocks and minerals leading to the formation of soils.

132. Oxidizing agent – The particular reactant of an oxidation reduction reaction that accepts electron/s from another reactant is the oxidizing agent; in other words, oxidizing agents are the electron acceptors. Further, by accepting the electron/s, the oxidizing agent gets reduced i.e. oxidizing agents during the course of causing oxidation get reduced. All oxidizing agents are not however, of equal potency and differ in the degree of their tendency to accept electrons. Greater the tendency of a substance to accept the electron, more powerful the substance is as an oxidizing agent. The order of the tendency of the four halogen atoms to accept electron is: F > Cl > Br > I. Therefore, among the four, F is the most powerful oxidizing agent followed by Cl, Br and I. Some of the widely used oxidizing agents in the laboratory are potassium permanganate, potassium dichromate, potassium iodate, potassium chlorate, concentrated nitric acid, concentrated sulphuric acid, hydrogen peroxide, ferric chloride, sodium hypochlorite, elemental chlorine, elemental iodine, and cerium sulphate.

133. Parent material of soils - The loose unconsolidated material obtained from the weathering of rocks and minerals which later on being subjected to further pedogenic processes forms the solum of soils is called the soil's 'parent material'. Parent material is thus the initial state of the soil system and is largely the C horizon of a soil in the description of soil profiles.

134. Particle density of soils - Particle density of soils is defined as the mass per unit volume of soil solids devoid of any pore space. Particle density of soils is almost a permanent character and is not influenced by addition of organic matter or tillage. Particle density is expressed as g per cm^3 or Mg per m^3.

135. Pedalfer - Podzolization and laterization form soils which belong to the Pedalfer group of soils (shifting of sesquioxides).

136. Pedocal - The soils of the low rainfall areas formed by calcification contain lime and gypsum very near the soil surface. Such soils with a calcic or gypsic endopedon are often referred to as Pedocals or Pedocal soils.

137. Pedology - Pedology is the study of soils purely as natural bodies with no emphasis for their immediate exploitation as a medium for plant growth and any other practical use.

138. Pedon – The term Pedon as used in Soil Taxonomy refers to a unit of soil volume for description and selection of a soil profile site and also collection of soil samples to represent a particular soil unit. A pedon is the smallest soil volume (area 1 m^2 to 10 m^2) which should describe a soil's horizons and variability present in the soil.

139. Pedoturbation - Pedoturbation is mixing of the constituents of soils taking place at varying degrees during soil formation.

140. Penning – The practice of keeping farm animals such as goats, sheep and cattle in enclosed fallow lands for days and nights after harvest of the last crop by way of giving them shelter and food and with the objective of enriching the soil fertility with the animals' urine and solid excreta is called penning.

141. Per cent Base Saturation – Per cent Base Saturation (PBS) of a soil is defined as the per centage of the cation exchange capacity of the soil occupied by the base forming cations viz. Ca^{2+}, Mg^{2+}, Na^{+} and K^{+}, both expressed in terms of m. e./100 g soil.

$$PBS = \left(\frac{\text{Exchangeable } [Ca^{2+}] + [Mg^{2+}] + [K^{+}] + [Na^{+}]}{\text{Cation Exchange Capacity}}\right) \times 100$$

PBS indicates the relative abundance of the base and acid forming cations (H^{+} and Al^{3+}) of soils and serves as an indicator for predicting the acidity or alkalinity of soils.

142. Permafrost – Permafrost refers to a ground below the surface that remains frozen (temperature 32^{0} F or 0^{0} C or still colder) for at least two years straight. Permafrosts are common in polar regions and in areas of high mountains. However, though the ground is frozen, the permafrost regions are not always covered in snow.

143. Permeability Index of irrigation water - Irrigation water with high sodium content may cause reductions in soil permeability. The effect of reduction in permeability is evaluated by the parameter called Permeability Index. Permeability Index is expressed as

Permeability Index = $\frac{[Na^{+}] + \sqrt{[HCO_3^{-}]}}{[Na^{+}] + [Ca^{2+}] + [Mg^{2+}]}$ x 100 where, the third bracket indicates the concentrations of the ions in terms of m. e./L.

144. Petrology – The branch of science concerned with the study of the different aspects of rocks such as their origin, structure, composition, properties, distribution and practical uses is called petrology.

145. pF - pF is defined as the negative logarithm to the base 10 of the numerical value of the soil moisture tension expressed in centimeters. The concept of pF as an index of the matric potential of soil moisture was first suggested by R. K. Schofield in 1935.

146. pH - pH is defined as the negative logarithm to the base 10 of the active H^+ ion concentration of an aqueous solution expressed in mole per L. The concept of pH as a mere number for expressing the acidity or alkalinity of a solution was introduced by S.P.L. Sørenson in 1909. The scale of pH ranges from 0 to 14. The pH of 1, 0.1 and 0.0001 N HCl solution are 0, 1 and 4, while the pH of 1, 0.1 and 0.0001 N NaOH solution are 14, 13 and 10, respectively.

147. Phosphate Potential – The concept of phosphate potential as a parameter of the availability of phosphorus to plants was introduced by Schofield (1955). In thermodynamic terms, phosphate potential is defined as the amount of work that must be carried out to move reversibly and isothermally an infinitesimally small amount of phosphate ion from a pool of phosphate at a specified location to the point under consideration at 1 atmospheric pressure. Schofield suggested that 'a labile pool of phosphate exists in soils from which the plant roots derive their phosphorus requirements'.

By measuring the pH and total P concentration in 0.01 M $CaCl_2$ solution, the solubility product (K_{sp}) of phosphate ions can be expressed as

$K_{sp} = [Ca^{2+}]^{1/2} [H_2PO_4^-]$. Upon taking negative logarithm (p = -log),

$$- \log K_{sp} = - \frac{1}{2} \log [Ca^{2+}] - \log [H_2PO_4^-] = - \frac{1}{2} \log [Ca^{2+}] + (- \log [H_2PO_4^-])$$

$$= \frac{1}{2} p\,[Ca^{2+}] + p\,[H_2PO_4^-] \quad (p = -\log).$$

This expression is known as the phosphate potential of soils. The equation for phosphate potential has thus its roots in the solubility product of phosphate ions. Since the availability of phosphate to plants is governed principally by the solubility of phosphates, the phosphate potential can justifiably be used to make fairly reliable predictions for the availability of phosphorus to plants. Low phosphate potential values of a soil suggests high availability of phosphorus to plants; the reverse is also true.

148. Phyllosilicates – Phyllosilicates, also referred to as sheet silicates, are characterized by their sheet like crystal structures. Phyllosilicates represent the class of polymeric silicate minerals in which the silicon-oxygen tetrahedral groups are linked by sharing three of every four oxygen atoms so as to form sheets of indefinite extent, in which the ratio of silicon to oxygen is 2:5, and in which some of the silicon atoms may be replaced by aluminium atoms. Micas, talc, serpentine, chlorite and clay minerals are phyllosilicates. Phyllosilicates display a platy or flaky crystal structure and near perfect cleavage.

149. Physical weathering – Physical weathering refers to the mechanical disintegration of bigger sized consolidated rocks and minerals into their smaller sized fragments, with however, no change in their chemical composition. Physical weathering is only a mechanical process and its operation is more conspicuous in very hot, very cold and very dry climates. Five principal agencies involved in physical weathering are temperature (heat), water, wind, waves, and glaciers (ice).

150. Plasticity - Plasticity is defined as the property which enables a system such as clay or soil to take up water to form a mass that can be deformed in any desirable shape and to maintain the shape after the deformation pressure is withdrawn. In soils, plasticity results from the plate-like nature of clay particles and the combined binding/lubricating effect of adsorbed water.

151. Plutonic rocks – Plutonic rocks are igneous rocks formed by the cooling and condensation of magma at considerable depth below the earth's surface. Cooling is thus slow, and the rocks are hard, massive and coarse grained. Granite, gabbro and syenite are important plutonic rocks. Plutonic rocks are also known as intrusive rocks.

152. Podzolization – Podzolization is an important soil forming process. In Podzolization (Russian words *Pod* and *zol* meaning under and ash) soil formation process, the oxides of iron and aluminium (sesquioxides) - often as complexes with humus - are washed down (eluviated) from upper layers and deposited at lower layers under conditions of high rainfall, cool humid climate, low pH, acidic parent material, and intense fungal activity. In such soils, A or E horizon is grey (due to silica), while the B or I horizon is dark coloured reddish or yellowish brown (due to sesquioxides).

153. Poiseuille's law - According to Poiseuille's law, the rate of flow of a liquid (Q) through a narrow tube is directly proportional to the fourth power of the radius (R) of the tube and to the pressure drop (ΔP) and is inversely related to the viscosity of the liquid (μ) and the length of the tube (L). Thus, $Q = \frac{\pi R^4 \Delta P}{8 \mu L}$. The equation shows that in a soil, water moves faster through water-filled pores of higher radii.

154. Potassium Adsorption Ratio – Potassium Adsorption Ratio (KAR) is a parameter for expressing the abundance of potassium content of soils and irrigation water. KAR is expressed as,

$$KAR = \frac{[K^+]}{\sqrt{\frac{[Ca^{2+}]+[Mg^{2+}]}{2}}}$$

In the above expression, all the cations are expressed in m. e./100 g soil or m. e./L of water. KAR of soils and irrigation water indicates the availability of potassium.

155. Potential Buffering Capacity of soils - Potential Buffering Capacity (PBC) of soils refers to the capacity of soils to restore the original pH of soils, despite the additions of acids and bases. Potential buffering capacity (PBC) of soils is directly related to the CEC of soils which in turn depends upon the clay content, nature of the clays and organic matter content of the soils

156. Poudrette - The concentrated organic manure obtained from night soil by its drying and deodourization by adding additives like charcoal and gypsum, but without any admixture with other waste materials, is called Poudrette.

157. Primary minerals – Primary minerals are those minerals which still retain the same chemical composition as they were in the original rocks. Quartz and feldspars are two examples of primary minerals.

158. Primary standard substance – If a standard solution of a substance can be prepared simply by dissolving an accurately weighed mass of the substance in a suitable solvent and then diluting to a known volume, then the substance in question is referred to as a primary standard substance. A substance to qualify as a primary standard substance must satisfy certain criteria such as availability in highly pure state, no hygroscopicity or efflorescence, high chemical stability, high molecular weight etc. Potassium hydrogen *ortho*phosphate, sodium carbonate, potassium dichromate, and ammonium oxalate are examples of primary standard acid, base, oxidizing agent and reducing agent, respectively. Sulphuric acid is not a primary standard acid for many of its defects. And so is the case with sodium hydroxide (base) and potassium permanganate (oxidizing agent).

159. Q/I Relationship – Quantity Intensity relationship, usually referred to as Q/I Relationship, of soils indicates the relative abundance and availability of free and exchangeable cations to plants in relation to the amount of the cations adsorbed by soil colloids. The concept of Q/I Relationship was originally proposed by Beckett for predicting the rate of exchange of K^+ in soils by Ca^{2+} or, by Ca^{2+} and Mg^{2+}. The Quantity factor (Q) refers to the concentration of K^+ in soils liberated by exchange with Ca^{2+}, while the Intensity factor (I) refers to the ratio of the activity of free K^+ ions to free Ca^{2+} ions or free K^+ ions to free $Ca^{2+} + Mg^{2+}$ ions *i.e.* $\frac{K^+}{Ca^{2+}}$ or, $\frac{K^+}{Ca^{2+} + Mg^{2+}}$

Therefore, the Q/I ratio relates the change in the concentration of K^+ in soils as affected by the activity ratio of $K^+/Ca^{2+} + Mg^{2+}$. The Q/I relationship is therefore formulated as

$Q/I = \frac{aK^+}{a(Ca^{2+} + Mg^{2+})}$ where '*a*' indicates the activity.

The Q/I ratio proves very useful for the study of short-term K^+ dynamics in soil solutions. The latter involves also the release of K^+ from nonexchangeable sites, and hence the ratio can be used as a measure of the availability of potassium to plant roots. The Q/I relationship can be expressed in the form of a curve in which the Quantity factor is represented by ΔK^+ (change in the concentration of K^+ in soil solution) on the Y axis and the Intensity factor is recorded on the abscissa representing the ion activity ratio $\frac{aK^+}{\sqrt{a(Ca^{2+} + Mg^{2+})}}$. The upper part of the curve is linear (attributed to nonspecific sites for K^+ ions related to planar surfaces) and the bottom part is curvilinear in regression (attributed to specific sites with high affinity for K^+ ions such as the edges of clay crystals). The slope of the linear portion of the curve is represented by the ratio aK/aAR^K ('*a*' is activity and AR is activity ratio). This ratio is called the Potential Buffering Capacity for potassium (PBC^K). This ratio is related to the CEC of the soil and is considered as an index of the soil's capacity to maintain the concentration K^+ at a constant value in the soil. A high PBC^K value states the potassium availability in the soil to be excellent. The reverse is also true implying the need for the application of potassic fertilizers.

160. Radioactivity – The term 'radioactivity' is used to describe the natural process in which the unstable atoms of certain elements undergo spontaneous disintegration, emitting both particulate matter and energy as they transform to different and more stable atoms. Radioactivity was discovered by Henri Becquerel (1896). The process of radioactive decay occurs because 'stability is the rule of nature' and thus unstable isotopes tend to transform to a more stable state. Radioactive decay is highly exergonic, statistically random and follows first-order kinetics that occurs with a small amount of mass being converted to energy. Since it is a first-order process, each radioactive species is characterized by its own half-life. Half-life is the length of time in which an initially very large number of such unstable atoms (rather, nuclei) get decayed to half of the original number. In radioactive decay, a relatively large amount of energy is liberated in each disintegration – typically about 1 million times more than the amount of energy liberated in an exothermic chemical reaction e. g. a few million electron volts (MeV) of energy per nucleus, compared to only a few electron volts (eV) of energy per atom or molecule. Since radioactive decay is a nuclear rather than an electronic phenomenon, its rate for a given radioactive species (radio isotope like ^{226}Ra or ^{239}U or ^{32}P or radionuclides) is not altered measurably by changes in temperature or pressure.

161. Radioactivity - Units – Three different units are in use for expressing the activity of radioactive materials as such or present in different substances such as plant, water and soil samples containing radioactive materials. They are Curie, Becquerel and Rutherford.

Curie – The unit of radioactivity called Curie (Ci) named after Marie Curie who discovered radium and polonium was in much use earlier. Originally, Curie represented the radioactivity corresponding to 1 g natural radium i.e. ^{226}Ra. Curie is now defined as the radioactivity equal to 3.7×10^{10} disintegrations (decays) per second (dps) i. e. 37 billion dps.

Thus, 1 Ci = 3.7 x 10^{10} dps. Curie is a very large unit and so milli Curie (1 mCi = 3.7 x 10^{7} dps), micro Curie (1 μCi = 3.7 x 10^{4} dps), nano Curie (1 nCi = 37 dps) and pico Curie (1 ρCi = 3.7 x 10^{-2} dps) are in greater use. The S. I. unit of radioactivity is Becquerel (Bq) named after Henri Becquerel who discovered radioactivity. One Bq is defined as the activity of a quantity of radioactive material in which one nucleus disintegrates (decays) in one second.

1 Bq = 1 disintegration (decay) per second (dps)

Interconversion of Ci and Bq – 1 Ci = 3.7 x 10^{10} dps = 3.7 x 10^{10} Bq

Therefore, 1 Bq = $\frac{1}{3.7 \times 10^{10}}$ Ci = 2.7 x 10^{-11} Ci

Or, 1 mCi = 3.7 x 10^{7}dps = 3.7 x 10^{7} Bq = 3.7 x 10 x 10^{6} Bq = 37 MBq (M for mega).

Or, 1 MBq = $\frac{1}{37}$ mCi = 0.027 mCi = 27 μCi

Rutherford (Rd) is the third unit of radioactivity (now obsolete).

1 Rd = 10^{6} dps = 10^{6} Bq = 1 MBq

Therefore, 1 Rd = 10^{6} Bq = 10^{6} x 2.7 x 10^{-11} Ci = 2.7 x 10^{-5} Ci = 27 μCi = 2.7 x 10^{4} n Ci = 2.7 x $10^{7}$$\rho$ Ci

162. Rainwater harvest – Rain water harvest is the practice of collection of rain water that runs off from roof tops, roads and open grounds for the purpose of storage for later use or for recharging of the depleting ground water table. The practice of rainwater harvest is gaining more and more significance in view of increased demand for water, climate change and prolonged dry spells of weather, unpredictable rainfall pattern, continued depletion of groundwater table, increasing salinity of the groundwater at shallow depths, and increasing pollution level of freshwater sources. Harvesting of rainwater can be practiced at wide ranging points such as individual homes, colonies, gated communities, multistoried apartments, institutions such as schools, colleges and universities, hostels, hospitals, clubs, slums, industries, hotels and anywhere with any scope for collection of the rain water. The practice of rainwater harvest helps (1) meeting the increasing demand of water, especially in areas where water is scarce, polluted and seasonally available, (2) improving the quality and quantity of groundwater, and (3) reducing the incidence of floods. Besides, rain water harvest offers a less expensive way of providing clean and pure water. Harvested rain water can be used for various purposes such as (i) drinking, food preparation and other direct human consumption, in which cases the harvested water needs to be filtered and sterilized, (ii) gardening and landscaping, (iii) swimming pools, (iv) livestock keeping, (v) aesthetics like water fountains, (vi) dishwashers, washing machines, bathrooms and toilets (grey water), and (vii) many more. A basic rainwater harvest system must have a way to collect the rainwater, usually along a slope, (2) a way to direct the water to a storage point taking advantage of slope and (3) a place to store the water. Many improved versions of rainwater harvest systems are now commercially available. Both Central and State governments have launched various schemes for rainwater conservation and its harvesting with people's participation. Two important ones among such successful programmes involving mass participation are the Jal Shakti Abhiyan. and Jal Jeevan Mission.

163. Ratio law – Equations describing cation exchange phenomenon state that the ratios of the amounts of cations adsorbed and cations free in solution are constants as explained by the Law of Mass Action and Gapon's equation i. e.

$$\frac{[Na^+]}{(Na^+)} = k\,\frac{[\sqrt{Ca^{2+}}]}{(\sqrt{Ca^{2+}})} = \frac{[Na^+]\,(\sqrt{Ca^{2+}})}{(Na^+)\,[\sqrt{Ca^{2+}}]}$$

In the above expression, [ ] denotes adsorbed cations and ( ) denotes cations free in solution. Therefore, the ratio of the cations in solution depends on the ratio of the cations adsorbed on the colloid surface. If the amounts of the cations adsorbed does not change significantly, or in other terms, the ratio of the adsorbed cations $\frac{[Na^+]}{[\sqrt{Ca^{2+}}]}$ remains unchanged, the ratio of the cations in solution $\frac{(Na^+)}{(\sqrt{Ca^{2+}})}$ also remains unchanged. Schofield in 1947 explained this activity ratio which is now known as Schofield's Ratio law. In simple terms, Schofield's Ratio law states that when the cations in solution (say, soil solution) are in equilibrium with a large number of exchangeable cations (say, on soil colloids), a change in the concentration of the solution shall not disturb the equilibrium, if the concentrations of all the monovalent ions are changed in one ratio, those of all the divalent ions in the square of that ratio, and those of all the trivalent ions in the cube of that ratio. Normally, the upper limit of the concentrations for which the ratio law applies is 0.1 M for monovalent ions like Na^+ and K^+, and 0.01 M for divalent ions like Ca^{2+} and Mg^{2+}. Ratio law, notwithstanding such limitations, does have high practical utility in governing exchange behaviour and nutrient availability in soils. However, it is more appropriate to use the activity of the ions in place of their concentrations.

164. Red soils – Red soils are red to yellowish in colour due to the coating of iron oxides on the surfaces of the soil peds. Red soils are formed from granites and gneiss either *in situ* or from the decomposed rock fragments washed down to the lower areas. Red soils are acidic to neutral in pH, normally poor in organic matter, N, P, and lime contents, and do not have high CEC and per cent base saturation. Red soils are moderately fertile. Surface crusting and hardening are the constraints of red soils. In Soil Taxonomy, the red soils belong to the soil Orders Alfisol, Inceptisol and Entisol. In India, the red soils occur in the states of Maharashtra, Odisha, Goa, North Eastern states, and the southern states of Tamil Nadu, Telangana and Karnataka.

165. Redox Potential – In oxidation reduction reactions, electrons are exchanged between a reducing agent and an oxidizing agent. The reactant that donates the electrons forms the reducing agent, while the reactant that accepts the electrons forms the oxidizing agent. On this score, the greater the tendency of an oxidizing agent to accept electrons, more powerful the oxidizing agent is. On the same token, the greater the tendency of a reducing agent to donate electrons, more powerful the reducing agent is. Redox potential (E) is just a quantitative measure of this tendency of an oxidizing agent to accept electrons or the tendency of a reducing agent to donate electrons. Redox potential is expressed in terms of volts (V). Both $KMnO_4$ and $K_2Cr_2O_7$ are oxidizing agents. However, between them, which one is more powerful and how much more powerful is indicated by their redox potentials (truly speaking, standard redox potentials, E^0). The oxidizing agent with a more +ve value of E^0 is the more powerful oxidizing agent. Values of E^0 of $KMnO_4$, $K_2Cr_2O_7$, and KIO_3 under standard conditions are +1.510 V, +1.330 V and + 1.195 V. Therefore, among the three, $KMnO_4$ is the most powerful oxidizing agent followed by $K_2Cr_2O_7$, and KIO_3.

166. Reducing agent - The particular reactant of an oxidation reduction reaction that donates its electron/s to be accepted by the oxidizing agent is the reducing agent; in other words, reducing agents are the electron donors. Further, by donating the electron/s, the reducing agent gets oxidized i. e. reducing agents during the course of causing reduction get oxidized. Reducing agents differ in the degree of their tendency to donate electrons. Greater the tendency of a substance to donate the electron, more powerful the substance is as a reducing agent. The order of the tendency of the three Group IA alkali atoms to donate electrons is: Li > Na > K. Therefore, among the three, Li is the most powerful reducing agent followed by Na and K. Some of the widely used reducing agents in the laboratory are sodium thiosulphate, sodium sulphite, hydrogen sulphide, ferrous sulphate, sodium arsenite, oxalic acid, sodium oxalate, ammonium oxalate, stannous chloride, potassium iodide, boron hydride, metallic zinc, and metallic iron.

167. Reduction - The term reduction in its narrowest sense refers to a chemical reaction in which hydrogen is added to or oxygen is removed from a substance. In more refined terms, in reduction, there is a gain of one or more electrons by an atom or ion or a molecule. The atom or ion or molecule is said to have undergone reduction i. e. be reduced. Decomposition of H_2O_2 to H_2O by removal of O_2, conversion of O_2 to H_2O by the addition of hydrogen, and flip of Fe^{3+} (ferric iron) to Fe^{2+} (ferrous iron) by gain of an electron are three simple examples of the process of reduction. Reduction is the reverse of the process of oxidation. In reality, reduction and oxidation take place simultaneously implying that while one reactant undergoes reduction, another one has to undergo oxidation. The reactant that undergoes reduction forms the oxidizing agent, while the reactant that undergoes oxidation forms the reducing agent. Further, there may be other reactants too in a reduction oxidation reaction which facilitate the processes of reduction and oxidation to proceed. Reduction is an important reaction in the chemical weathering of rocks and minerals leading to the formation of soils.

168. Regolith - The loose unconsolidated material overlying the rocks is called the Regolith. Regolith at a place, however, might have been formed from the underlying rocks (*in situ*) or might have been formed due to its transportation and deposition over the bedrocks by the agencies of nature such as the ice, water, wind and gravity.

169. Regosol – Regosols are intrazonal soils characterized by shallow and medium to fine textured unconsolidated parent materials that may be of alluvial origin. Regosols lack a significant horizon differentiation due to dry or cold climatic conditions. Regosols are largely similar to the soils of the Order Entisol.

170. Regur – The fertile black soils formed by the weathering of basaltic igneous rocks and widespread in Deccan plateau and Telangana are locally called Regurs.

171. Relief as a soil forming factor - Relief refers to the configuration of the land surface and includes features of the land like slope, elevation and position of the landscape. In soil formation, relief often acts as a climate modifier and can hasten or delay the soil formation process. Relief influences its role by controlling runoff, drainage, soil erosion and by creating a new, but distinctly different, microclimate. It may be mentioned in this context that a group of soils of about the same age and formed under the same climatic conditions and from the same parent material, but under different relief, is called a Catena. Catena signifies the role of relief in soil formation.

172. Reserve acidity of soils - Reserve or residual acidity of a soil is defined as the soil's acidity due to the presence of H^+, Al^{3+} and aluminium hydroxyl ions [$Al(OH)^{2+}$ and $Al(OH)_2^+$] present in nonexchangeable forms in silicate clays and organic matter. The magnitude of reserve acidity of soils is far higher than that of active acidity of soils (10^3 to 10^4 times or still higher).

173. Residual Sodium Bicarbonate of irrigation water – A property of irrigation water that must be considered for judging its suitability is its Residual Sodium Bicarbonate (RSBC). RSBC of irrigation water is important where the bicarbonates are more dominantly present than the carbonates. In such cases, RSC, the more commonly used characteristic, is better expressed as RSBC. RSBC of water is expressed as follows.

$$RSBC = [HCO_3^-] - [Ca^{2+}]$$

The two ions in the above expression are expressed in m. e./L. RSBC thus indicates the difference between the concentrations of the bicarbonates and the much important divalent cation Ca^{2+}. Stated otherwise, RSBC value compares the presence of bicarbonates *vis a vis* that of calcium on equivalent basis.

174. Residual Sodium Carbonate of irrigation water - Residual Sodium Carbonate (RSC) of irrigation water is a parameter for predicting the sodium or alkalinity hazard of the water. RSC of water is expressed as

$$RSC = ([CO_3^{2-}] + [HCO_3^-]) - ([Ca^{2+}] + [Mg^{2+}])$$

The four ions in the above expression are expressed in m. e./L. RSC is thus the sum of the carbonates and bicarbonates minus the sum of the two favourable divalent cations viz. Ca^{2+} and Mg^{2+}. As the value of RSC increases, the possible sodium hazard of the irrigation water too increases, because the water adds more carbonates and bicarbonates than the divalent cations that help improve the soil physical properties. RSC value is an indicator of the potentials for the precipitation of Ca^{2+} and Mg^{2+} at the soil surface and thus the removal of Ca^{2+} and Mg^{2+} from the soil solution. RSC values of irrigation water of < 1.25, 1.25 to 2.5, and > 2.5 indicate the water to be safe, marginally safe and usually unsuitable for the purpose of irrigation, respectively.

175. Rocks – Rocks are naturally occurring substances and are aggregates of two or more rock forming minerals in varying proportions. Since they have many constituents in undefined proportions, they do not have definite physical and chemical properties. Granite and basalt are two examples of rocks.

176. Rubification – Rubification refers to the process in which soils develop a distinct red to yellow colour due to the formation of hematite as a coating agent on clays during the process of intense weathering in warm humid climates.

177. Runoff – Runoff refers to the water flow or precipitation that does not get absorbed (infiltrated) into the soil or does not evaporate and therefore makes its way along a slope from the ground surface into places that collect the water. Runoff causes soil erosion and also carries chemicals and substances present on the ground surface along with it to the rivers, lakes, ponds or low lying areas where the water so moved ends up. Runoff thus causes silting of water bodies and also water pollution.

178. Salinization of soils - The process in which soils accumulate soluble salts such as the chlorides, sulphate etc. of calcium, magnesium, potassium and sodium in high amounts in the upper soil layers, sometimes in the form of a distinct horizon (known as salic horizon), is called salinization. The soils so formed are called saline soils (Solonchaks). Saline soils are characterized by EC > 4 dS/m, ESP < 15 and pH < 8.5 (usually 7 to 8.5). Salinization is of common occurrence in (1) dry areas of low rainfall where potential evapotranspiration is more than the quantum of rainfall, (2) low lying areas/depressions (such as the deltas and flood plains) where water with salts are likely to stand, (3) areas of imperfect or poor drainage restricting the salts to leach down, (4) alluvial deposits along the sea coast, (5) areas where the ground water rich in dissolved salts (such as the sulphates and chlorides of Ca, Mg, K and Na) is under the direct influence of excessive evaporation and transpiration, (6) inundation of land areas by sea water and (7) use of water rich in salts (saline water) for irrigation. Yet, one more major reason for the formation of saline soils is the fluctuating ground water table. When there is a capillary rise of ground water rich in salts, the salts also rise upwards. Later when the water recedes down or dries up, the salts are left behind at a soil depth depending upon the capillary rise. If the accumulation of the salts is at a depth very close to the soil surface, the salts can be recognized by their white appearance. Saline soils are therefore known as white alkali soils also. A major way for the reclamation of saline soils is the leaching down of the salts of the surface layers by using good quality water followed by good subsurface drainage. Leaching Requirement (LR) is an important aspect for the reclamation of saline soils.

179. Salt Affected Soils - Salt Affected Soils refer primarily to saline, saline alkali and alkali (sodic) soils. These soils are known as problem soils. Chlorides and sulphates of calcium, magnesium, potassium and sodium are the principal salts in these soils. These soils are more common in (i) dry areas where rainfall proves to be inadequate to leach down the deposited salts, (ii) areas where the parent material is rich in sodium, (iii) areas close to coast line, and (iv) areas where the irrigation water rich in salts including sodium salts does change the normal soils to saline or saline sodic or sodic soils over a period of time. Soil properties which characterize these soils are pH, EC and ESP. Saline soils have pH < 8.5, EC > 4 dS/m and ESP < 15; saline alkali soils have pH > 8.5, EC > 4 dS/m and ESP > 15 and alkali soils have pH > 8.5, EC < 4 dS/m and ESP > 15. Saline soils are referred as white alkali soils because of a white encrustation due to salt deposit on the soil surface, while alkali or sodic soils are referred as black alkali soils due to the dissolution of organic matter by high pH, thus imparting a black colour to the soil. Saline soils are reclamed by leaching with good quality water, while saline sodic and sodic soils are reclamed by using gypsum as the soil amendment. In Soil Taxonomy, salt affected soils belong to the soil Orders Aridisol, Inceptisol, Entisol, Alfisol and Vertisol.

The problem of salt-affected soils is a growing concern, more so in arid regions. In India, these soils cover an area of about 10 million ha and are widespread in the states of Punjab, Haryana, Gujarat, Rajasthan, Maharashtra and Karnataka.

180. Salt Index of fertilizers – Salt Index of fertilizers refers to a number which is the ratio of the increase in osmotic pressure produced by a given weight of the fertilizer to that (increase in osmotic pressure) produced by the same weight of sodium nitrate ($NaNO_3$), where the latter is assigned a value of 100. The choice of $NaNO_3$ is largely because of its 100 per cent solubility. Salt Index of fertilizers is indicative of the hazard of salt build-up in soils because of fertilizer application. Salt Index of sulphate of potash (SoP) and muriate of potash (MoP) are 42.6 and 120.1, respectively. MoP thus contributes more to the increase in salinity level of soils.

181. Salt Index of irrigation water - Salt Index of irrigation water is an experimental parameter that expresses the relationship among the concentrations of sodium, calcium and $CaCO_3$. The Salt Index so stated is calculated as follows.

Salt index = (Total Na^+ – 24.5) – [(total Ca^{2+}– Ca in $CaCO_3$) x 4.85]

In the above expression, total Ca^{2+} is Ca^{2+} + Mg^{2+} and the ions are expressed mg L^{-1}. Salt Index is an indirect water quality parameter for predicting the sodium hazard of irrigation water. If the Salt Index is negative, the water is well suited for irrigation and *vice versa*.

182. Saltation (in wind erosion) – Saltation is the most important of the three processes of wind erosion responsible for the transportation of soil particles after their detachment from soil mass. The other two are surface creep and suspension. Saltation causes movement of medium sized soil particles (0.05 to 0.5 mm diameter, light enough and hence, easy to be lifted) by a short series of bounces, usually not rising above 30 cm, along the ground surface. In wind erosion, saltation accounts for 50 to 75% of the total movement of the dislodged soil particles depending upon weather conditions, the rest being accounted by surface creep and suspension.

183. Scarification – The process of weakening/softening/opening/ altering the hard seed coat of the seeds of some plants to facilitate their quick germination is called scarification. Scarification is done mechanically or thermally or chemically to modify the seed coat that is otherwise largely impervious to the entry of water and gases. For the seeds of dhaincha, a green manuring plant, scarification is done mechanically by light pounding with sand.

184. Secondary minerals – Secondary minerals are those minerals which have been formed from the primary minerals due to the changes in them suffered during the weathering of rocks and minerals. Secondary minerals are thus altered minerals and do not retain their original mineralogical properties. Clay minerals are examples of secondary minerals.

185. Sedimentary rocks - Sedimentary rocks are formed from the preexisting igneous and metamorphic rocks and various minerals due to the effects of weathering, transportation, sedimentation, diagenesis and cementation. Sandstone, shale, and limestone are examples of sedimentary rocks.

186. Sewage – Sewage refers to the wastewater produced from a community of people and is made up of (1) grey water originating from sinks, bath tubs, showers, dishwashers, washing machines etc, (2) black water that has been used to flush toilets along with human wastes and soaps, (3) shampoos and detergents, and (4) toilet papers. Sewage passes into the sewerage system.

187. Sewerage - Sewerage is the pipe system that carries the sewage for its disposal or purification, reuse and recycling.

188. Shelter belt – Shelterbelt is a wide belt of trees, shrubs and grasses planted in rows at right angle to the direction of the prevailing winds to deflect the air current, reduce wind speed and extend a general protection to the areas against the desiccating effect of the hot winds by creating a microclimate. A shelterbelt is a kind of agroforestry to reduce soil loss by wind erosion, provide protection to crops and livestock from the wind, conserve soil water, encourage wildlife and create a rich carbon bank.

The difference between a shelterbelt and a windbreak, the former's close associate, is very thin. Windbreak consisting of two to three rows of plants is a protective planting against a garden, farm or field, while shelterbelt consisting of multiple rows of plants along fence lines are intended for protection of animals and crops. Both however, reduce wind erosion, conserve soil moisture and provide a rich carbon bank as their major mandates.

189. Slickensides – Slickensides is a characteristic feature of soils that exhibit swell shrink property due to the presence of expanding type clay minerals. Slickensides are prominently noticed in black soils of the Order Vertisol. Slickensides – polished and grooved surfaces, generally exceeding 5 cm – are produced when one soil mass slides past another. Some among them occur at a lower boundary of a slip surface where a mass of soil moves downward on a relatively steep slope.

190. Sludge - Sludge is the sediment that settles down in the activated sludge process. It is a dark colored powdery material, and does have good manurial value.

191. Sodium Adsorption Ratio - Sodium Adsorption Ratio (SAR) is a numerical value for expressing the alkalinity of soils and sodium hazard of irrigation water. SAR is expressed as

$$SAR = \frac{[Na^+]}{\sqrt{\frac{[Ca^{2+}] + [Mg^{2+}]}{2}}}$$

In the above expression, all the cations are expressed in m. e./100 g soil or m. e./L of water.

SAR is an important characteristic of Salt Affected Soils. SAR takes into account the fact that the adverse effect of Na^+ is moderated by the presence of Ca^{2+} and Mg^{2+} ions. When the SAR rises above 12 to 15, serious soil physical problems arise and plants find difficulty in absorbing water. A SAR value of > 13 is one of the criteria for a soil to be categorized as an alkali soil.

Based on SAR, irrigation water is classified into different classes for their safe, limited, restricted and prohibited use. Irrigation water with SAR values of upto 10, 10 to 18, 18 to 26 and > 26 are considered low, medium, high and very high sodium hazard irrigation water, respectively.

192. Soil Amendments - Soil amendments are substances that are to be applied to problem soils for their reclamation. Limestone, dolomite and basic slag are some of the commonly used soil amendments for the reclamation of acid soils, while gypsum, pyrites and elemental sulfur are some of the commonly used soil amendments for the reclamation of alkali soils.

193. Soil Conditioners – Soil Conditioners are synthetic organic macromolecules added to soils for the aggregation of soil separates leading to favourable soil structure. Important among them are polyvinyl alcohol (PVA), polyethylene glycol (PEG), polyacrylamide (PAM), polyvinyl acetate (PVAc) and polyurethane.

194. Soil Consistence - Soil Consistence is defined as the manifestation of the physical forces of cohesion and adhesion acting within the soil at various moisture constants. Soil consistence is a dynamic soil physical property varying with soil moisture contents and applied physical stress. There are thus different consistencies of dry, moist and wet soils. These designated consistencies are sticky and plastic with further subdivisions for wet soils; loose, very friable, friable, firm, very firm and extremely firm for moist soils; and loose, soft, slightly hard, hard and extremely hard for dry soils

195. Soil Crusting – Soil Crusting refers to the deterioration of soil structure, where the natural soil aggregates break away and then disperse due to the impact of the fast falling rain drops, followed by the radiant energy of the sun. Soil crusting increases bulk density, decreases soil aeration and water movement, and restricts root penetration and seedling emergence. Soil factors associated with crusting are low organic matter, high silt and high exchangeable sodium contents.

196. Soil Horizon - A soil horizon is a layer of soil approximately parallel to the land surface. A soil horizon can be distinguished from the adjacent ones by distinctive soil properties like soil colour, texture, structure, organic matter content, mineral composition etc. A soil profile is the soil's vertical exposure showing the soil horizons. Master horizons of soils designated as O, A, E, B and C horizons are used in characterization of soils.

197. Soil Permeability - The ease with which water passes through a bulk mass of soil or a layer of soil is called the permeability of the soil. Obviously, the permeability of the surface soil is more than that of the subsurface soil. The presence of an impervious layer reduces the permeability of the soil. Permeability of soils is improved by growing deep rooted crops.

198. Soil Profile - A vertical section of a soil that opens up the soil horizons running upto the parent material is called the profile of the soil. A soil profile is thus a vertical exposure of the soil showing all the distinct soil horizons. Horizons are layers of soil approximately parallel to the land surface and one of them can be distinguished from the adjacent one by distinctive soil properties like soil colour, texture, structure, organic matter content, mineral composition etc. A soil profile with its visibly distinguishable soil horizons is a historical record of the soil forming processes and is of great use in soil classification and Soil Taxonomy.
A soil profile does have five master horizons designated by five capital letters viz. O (made up of organic materials), A (uppermost mineral horizon), E (eluvial horizon), B (illuvial horizon) and C (unconsolidated material underlying the solum). Subordinate layers or distinctions within the master horizons are designated by lower case letters or numerals e. g. O_i, O_e, O_a or A_1, A_2, and A_3.

199. Soil Structure - Soil Structure may be defined as (1) the size, shape, and arrangement of the soil particles and aggregates, (2) the size, shape, and arrangement of the voids or spaces separating the particles, and (3) the combination of the aggregates and voids into various types of structure. Soil structure is classified in different Types (based on shape and arrangement), Classes (based on size) and Grades (based on distinctness and durability).

200. Soil Taxonomy – Soil classification systems followed in the early period of soil study suffered from many defects. For instance, there was too much emphasis on unseeable features like genesis and climate e. g. tundra soils (very cold region soils) and chernozems. In some instances, soil colour was the only criterion for naming a soil e. g. black soils, red soils and brown soils. To overcome these defects of soil classification, the Soil Survey staff of USDA initiated works in 1951 to arrive at a more appropriate, comprehensive, and meaningful system of soil classification. Each year starting from 1953, an approximation was published and tasted for its acceptance. In 1960, a Comprehensive System of Soil Classification, now known as the '7th Approximation' was published in appropriate domains for deliberations. Classification of soils suggested in the 7th Approximation was widely accepted by the majority concerned. Subsequently, several more approximations were published and debated for acceptance. This ongoing and continuously evolving system of classification is now known as Soil Taxonomy. The system is now followed in most countries including India. In Soil Taxonomy (i) soil classification is based on soil properties (colour, pH, texture etc.) as they are seen in the field today and their is no importance on any unseen or unseeable or unobservable attributes like geology, parent material etc., (ii) nomenclature uses coined words derived from several languages, mostly from Greek and Latin. These words may initially sound unfamiliar and even awkward, but become very acceptable after initial reluctance, if any, and (iii) additions, deletions, and modifications based on difficulties, advantages, experiences, and deliberations are continuously implemented. As on now, all soils are classified in twelve Orders under which there are five more subcategories. The hierarchy is thus: Order → Suborder → Great Group → Subgroup → Family → Series → (Phase).

201. Soil Taxonomy - Categories – In Soil Taxonomy, there are six categories for classification of soils. The category at the highest level is the Order. There are twelve soil Orders viz. Alfisols, Andisols, Aridisols, Entisols, Gelisols, Histosols, Inceptisols, Mollisols, Oxisols, Spodosols, Ultisols and Vertisols. The Order is followed by five lower categories viz. Suborder, Great Group, Subgroup, Family and Series. The hierarchy of the system of soil classification in Soil Taxonomy now in vogue is thus: Order → Suborder → Great Group → Subgroup → Family → Series. The bottommost category is thus the Series. As one goes down in the system, specificity of the soil properties increases. The six categories are often grouped into two broad classes viz. higher categories which include Order, Suborder, and Great Group, and lower categories which include the remaining three viz. Subgroup, Family and Series. Often, the Series is followed by Phase which indicates a very important local property such as rocky, stony and slopy. Phase is however, not a taxonomic category, but is of great use in describing a soil and also in soil cartography.

202. Soil Taxonomy - Nomenclature – Nomenclature of Soil Taxonomy uses coined words from several ancient languages like Greek, Latin and also some other modern languages (German, French, Spanish etc.). Such coined worded names of the soils sound very euphonic, indicate few prominent soil properties and become easily rememberable. Initially, the coined worded names may sound a bit strange and awkward, but become very understandable, acceptable, and easily rememberable after some time. Further, each part of the coined names of the soils conveys one or more central character/s of the soils. Thus, if the Order of the soil is Ultisol, the soil must be very matured (*ultus*, a Latin word meaning ultimate; ultimate here means matured; *sol* means solum or soil). Or, if the Order of the soil is Entisol, the soil must be very young (*Ent* in the context is derived from the English word Recent, meaning young; *sol* means solum or soil). In naming a Suborder, the formative words of the names of the Orders are added to a significant character of the soil in question. These formative words for the twelve Orders are *alf, and, id, ent, el, ist, ept, oll, ox, od, ult* and *ert* for Alfisols, Andisols, Aridisols, Entisols, Gelisols, Histosols, Inceptisols, Mollisols, Oxisols, Spodosols, Ultisols and Vertisols, respectively. Accordingly, if a particular Ultisol soil is wet, then the Suborder of the soil is Aquult (*aqu* of aqua or water has been joined with the formative word or element *Ult* of Ultisol to make the conjoined word Aquult). Similarly, if a particular Entisol soil is wet, then the Suborder of the soil is Aquent. Similarly, we may have Aqualf, Aquoll, Aquod, Aquert and Aquept (but not Aquid since an Aridisol or dry soil cannot be wet). Reasoning and logic continue for the nomenclature of the other categories under the Suborders viz. Aquult and

Aquent. Thus, for a soil Suborder Aquent, the Great Group and Subgroup may be Cryaquent (*cryic* soil temperature regime) and Typic Cryaquent (typic means ideal), respectively. One thus moves from Order to Suborder to Great Group to Subgroup as Entisol→ Aquent → Cryaquent → Typic Cryaquent. It is advisable to refer a book on soil taxonomy for an in-depth study of the nomenclature of soils right upto the level of Series and Phase.

203. Soil Taxonomy - Soil Moisture Regimes – Soil Moisture Regimes (SMR) help framing the meaningful name of a category of a soil in Soil Taxonomy. Soil moisture regimes are defined in terms of the level of ground water and seasonal presence or absence of water held at a tension of less than 1500 kPa in the moisture control section as stipulated in Soil Taxonomy. There are five SMRs viz. Aquic (Latin *aquic* meaning water), Udic (Latin *udas* meaning humid), Ustic (Latin *ustus* meaning burnt or dryness), Xeric (Latin *xeros* meaning dry), and Aridic & torric (Latin *aridus* meaning dry and *torridus* meaning hot and dry). It is advisable to refer a book on soil taxonomy for more details of the individual SMRs. Aquod (Spodosol), Udult (Ultisol), Ustert (Vertisol), and Xeroll (Mollisol) are Suborders in Soil Taxonomy signifying aquic, udic, ustic, and xeric SMR, respectively. Torriorthent is a Great Group under the Order Entisol specifying torric SMR.

204. Soil Taxonomy - Soil Temperature Regimes – Soil Temperature Regimes (STR) help framing the meaningful name of a category of a soil in Soil Taxonomy. Soil temperature regimes are defined in terms of the mean annual soil temperature (MAST) measured at 50 cm depth and seasonal variation between mean summer soil temperature (MSST) and mean winter soil temperature (MWST). There are six STRs viz. gelic (L. *gelare*, to freeze; also, subgelic, pergelic and hypergelic), cryic (Gr. *kryos*, coldness), frigid, mesic, thermic and hyperthermic. For instance, Mesic STR refers to MAST of 8^0 C or higher, but lower than 15^0C, and the difference between MSST and MWST is 6^0 C or more at specified depth points. The prefix iso (isofrigid, isomesic etc.) indicates the difference between MSST and MWST is less than 6^0 C. Cryaqualf is the name of a Great Group signifying cryic STR (and aquic SMR). It is advisable to refer a book on soil taxonomy for more details of the individual STRs.

205. Soil Texture - Soil Texture is defined as the relative proportion of the soil separates of various sizes (Soil separates/Mechanical fractions) such as sand, silt, and clay. There are different textural classes of soils based on the relative abundance of sand, silt, and clay. Soil texture is almost a permanent physical property of soils and changes, if at all, get into only after a long period of time. Knowledge of the soil texture is very important in assessing the edaphological value of the land. Land capability and methods of soil management are determined primarily by soil texture.

206. Soils of India – India is a country endowed with great variations in geology, climate, relief features, landforms, vegetation, vast land area (328.5 m ha) and three distinct physiographic zones viz. peninsula (Deccan plateau and south of the Vindhyas), extra peninsula (mountain region of the Himalayas) and Indo Gangetic plains (which separate the two). The country does justifiably exhibit diverse kinds of soils. In terms of Soil Taxonomy, all the soil Orders, except Andisols, and Spodosols, are found in India. For common usage, Indian soils are categorized into eight types viz. alluvial, black, red, laterite and lateritic, desert/arid, forest and hill, peaty and marshy, and salt affected soils. Among these soils, the alluvial soils are the most widespread and account for nearly 40% of the land area and are very fertile too followed by black and red soils. Salt affected soils include saline, saline alkali and alkali soils. They are the problem soils. In India, the area under such soils is on steady rise due to lack of soil management – a subject of much concern.

207. Solonchak soils – Solonchak soils constitute an important Soil Great Group under Intrazonal Soils. Common accepted international names for Solonchak soils are saline and salt-affected (halomorphic soils). These soils are widespread in arid and semiarid climatic zones and in coastal regions of all climates. Solonchak soils are characterized by high soluble salt concentration within 30 cm of the land surface and by the absence of distinct subsurface horizonation, except for accumulation of gypsum, sodium or calcium carbonate or layers showing the effects of waterlogging. In modern Soil Taxonomy, majority Solonchak soils come under the Order Aridisol.

208. Solonetz soils – Solonetz soils constitute an important Soil Great Group under Intrazonal Soils. Common accepted international names for Solonetz soils are alkali or sodic soils. Solonetz soils are characterized by an accumulation of sodium salts in the surface zone and a strongly structured clay layer with adsorbed sodium ions on the exchange complexes in the surface and subsurface depths. In modern Soil Taxonomy, majority Solonetz soils come under the Order Aridisol and Molloisol with a natric endopedon. A warm climate is not a mandatory requirement for the formation Solonetz soils, unlike the Solonchak soils.

209. Solubility Product - The solubility product (K_{sp}) of a sparingly soluble salt at a particular temperature is defined as the product of the molar concentrations of the constituent ions in its saturated solution, each concentration term being raised to a power representing the number of the ions of that constituent ion produced by the dissociation of one molecule of the salt. K_{sp} of AgCl and $Ca_3(PO_4)_2$ are therefore $[Ag^+]$ x $[Cl^-]$ and $[Ca^{2+}]^3$ x $[(PO_4)^{3-}]^2$, respectively. The solubility product is an important aspect in governing the solubility and precipitation of salts in heterogenous soil systems and does have much practical and commercial applications too such as the manufacture of pure table salt from sea water and manufacture of baking soda.

210. Solum - The horizons of a soil profile above the parent material viz. A, E and B horizons are collectively called the 'solum' of the soil The solum of a soil thus is made up of a set of horizons that are related through the same cycle of pedogenic processes. The solum of a soil may be conceived also as the regolith minus the C horizon or the parent material of the soil.

211. Spectrophotometry - Spectrophotometry is an analytical tool based upon the interaction of an incident electromagnetic radiation with matter. In spectrophotometry, electromagnetic radiation (visible or UV or IR generated from an appropriate source) in the form of a narrow range or even individual wavelengths (monochromatic radiation) is passed through the sample solution and then the intensity of the radiation absorbed by the solution, or the intensity of the radiation transmitted through the solution is measured by suitable devices. The extent of absorption or transmission is related to the concentration of the substance in the solution that is to be determined. The interaction of the electromagnetic radiation with matter is thus made quantitative and this quantification makes spectrophotometry a very simple, sensitive, versatile and rugged analytical tool for the determination of a wide ranging substances e. g. medicines, pesticides, fertilizers, forensic samples, biochemicals and so on. Spectrophotometry uses a spectrometer which causes dispersion of the incident radiation into narrow range or individual wavelengths and a photometer which measures the intensity of the radiation. Depending upon the wavelength region of operation, spectrophotometry may be visible or UV and or IR spectrophotometry. Further, depending upon the measurement of absorption or emission of radiation, spectrophotometry may be absorption spectrophotometry e. g. colorimetry or emission spectrophotometry e. g. flame photometry

212. Standard solution – A solution of known concentration is called a 'standard solution'. A 0.1 N HCl solution is a standard solution because its concentration is known (0.1 N). Similarly, 10 ppm iron and 1% sucrose solution are standard solutions of iron and sucrose. Standard solutions are of different grades. A solution prepared from a primary standard substance is a primary standard solution. A solution the concentration of which has been determined by stoichiometric titration against a primary standard solution is called a secondary standard solution. Similarly, a solution the concentration of which has been determined by stoichiometric titration against a secondary standard solution is called a tertiary standard solution.

213. Sticky Point – Sticky Point, a particular consistence of wet soils, represents the soil consistence at which the soil moisture content is such that the soil does not stick to a foreign object, say a steel or plastic spatula. Sticky point represents the maximum moisture content for a soil to remain friable.

214. Stokes' law - Stokes' law describes the quantitative relationship between the terminal velocity of a falling spherical particle (v) such as a soil particle in a viscous fluid medium with the radius (r) and density of the particle (D_p), and density and viscosity of the fluid (D_f and η). Thus,

$$v = \frac{2\,(D_p - D_f)gr^2}{9\,\eta},$$

In the above expression, '*g*' represents the acceleration due to gravity.

215. Sullage – Sullage refers to the waste water drained out from kitchens, bathrooms, and drainage water flowing through the open street/roadside channels/drains. Sullage however, does not include any liquid or solid excreta emanating from the toilets.

216. Surface creep (in wind erosion) – Surface creep is one of the three processes involved in wind erosion for the transportation of soil particles after their detachment from the soil mass. The other two are saltation and suspension. Surface creep causes movement of soil particles (0.5 to 2 mm diameter, larger than those in the case of saltation) by a few meters by means of rolling and sliding along the ground surface. Surface creep is encouraged by saltation as the bouncing particles hit the large aggregates and speed up their movement along the surface. In wind erosion, Surface creep accounts for 5 to 15% of the total movement of the dislodged soil particles, the rest being accounted by saltation and suspension.

217. Suspension (in wind erosion) – Suspension is one of the three processes involved in wind erosion for the transportation of soil particles after their detachment from the soil mass. The other two are saltation and surface creep. In suspension, the dust particles upto fine sand size (soil particles upto 0.2 mm diameter) are moved both parallel to the ground surface and upward. The turbulent action of the wind moves the particles hundreds of miles horizontally and few miles upward in the atmosphere, which later return to the ground when wind velocity subsides or are washed by precipitation when it occurs. In wind erosion, suspension is thus a long-distance transportation of the soil particles, and accounts for 10 to 40% of the total movement of the dislodged soil particles (average 15%), the rest being accounted by saltation and surface creep.

218. Suspension effect (in soil pH measurement) – Measurement of soil pH recorded by a pH meter is affected by several factors. An important one among them is the effect of the soil suspension known as the 'suspension effect'. Considerable variation is noticed between the recorded pH of the supernatant and sediment of the soil suspension. If the soil suspension is allowed to settle down, the pH as measured in the suspension liquid is often higher than that in the sediment layer. The suspension effect is influenced by the extent to which the electrodes of the pH meter encounter clay and humus particles and the soil CO_2 is in equilibrium with its atmospheric concentrations, and the magnitude of the liquid junction effects. The effect of 'suspension effect' is minimized by measuring the soil pH using 0.01 M $CaCl_2$ or 1 M KCl solution instead of water. The use of 0.01 M $CaCl_2$ or 1 M KCl solution in place of water does have the additional advantages of (i) decreasing the effect of the junction potential of the calomel reference electrode, (ii) equalizing the salt content of soils, and (iii) preventing dispersion of the soil. Because of suspension effect, most soil analysts while reporting soil pH report the solution used for making the suspension and soil:solution dilution too e. g. Soil pH_{KCl} (1:2.5) – X, where X is the recorded pH that is to be reported

219. Sward – Sward is a dominantly rhizogenous surface soil horizon of grasslands and includes Mollic epipedon or Ap horizon formed by the practices of cultivation.

220. Topography – Topography is a synonym for relief. Topography refers to the configuration of the land surface and includes features of the land like slope, elevation and position of the landscape. In soil formation, topography often acts as a climate modifier and can hasten or delay the soil formation process. Topography influences its role by controlling runoff, drainage, soil erosion and by creating a new, but distinctly different, microclimate.

221. Tyndall effect - Tyndall effect is an optical property of a colloidal solution reported by J. Tyndall. If a beam of light is allowed to fall on a colloidal solution, the colloidal particles of the solution scatter the incident light and the path of the light can be easily detected even by human eye. The phenomenon is called Tyndall effect. Tyndall effect is caused by the scattering of light by the particles of the dispersed phase of the colloidal solution. Tyndall effect distinguishes a colloidal solution from a true solution.

222. Universal Soil Loss Equation - Universal Soil Loss Equation (USLE) is an empirical equation for predicting annual soil loss caused by water erosion. The equation was formulated by W. H. Wischmeier and D. D. Smith of USDA in 1950s from the results secured from a large mass of field data, especially erosion plots and experiments. The equation predicts soil loss as

$$A = R.\ K.\ L.\ S.\ C.\ P$$

where, A = computed soil loss per unit area in tons per acre or ha per year, R = is the rainfall-runoff factor of the geographical location, K = soil erodibility factor, L = slope length factor, S = slope gradient (steepness) factor, C = cropping management factor, and P = erosion control practices factor. By making use of the equation, personnel involved in soil conservation endeavour reducing soil loss by water erosion by (1) decreasing L (slope length) and S (slope steepness), (2) modifying C (cropping management factor) by practices like direct drilling, stubble mulching, strip cropping and alley farming, and (3) modifying P (soil management or erosion control practice factor by adopting practices like contour banks, terraces etc.). Since, L, S, C and P can be modified, the equation may be simplified in the following form.

$$A = R.\ K.\ (L.\ S.\ C.\ P) = R.\ K.\ X$$

The equation is in much use for assessing the effectiveness of soil conservation programmes. The equation however, does not apply to soil loss in extreme situations like that from large gullies and non-agricultural lands.

223. Value of colours - Value or brilliance of colours expresses apparent lightness as compared to absolute white. Precisely, it refers to the relative brightness or darkness of a colour within a scale of 0 to 10 as compared to absolute white. It in essence refers to the gradations from white to black (lightness or darkness). In the numerical notation 2.5 YR 5/6 for a soil colour read from Munsell Soil Colour chart, the digit 5 indicates the 'value' of the soil colour.

224. Vermicomposting - Vermicomposting is a method of composting in which the organic refuse is acted upon by earthworms as the composting agents. The product is called vermicompost. In vermicomposting, the earthworms serve as the multipurpose mixers, crushers, degraders and aerators. Vermicomposting is a process of regeneration of wealth from wastes that are otherwise simply dumped, often injudiciously adding to the pollution of the environment. In addition to the specialized earthworms called vermiworms such as *Eisenia foetida* and *Eudrilus eugeniae*, the local ones can also be used (by collecting them in small numbers and then multiplying them). Earthworms are very sensitive to sunlight and rains. Vermicomposting therefore, is to be practiced only under proper shade. It is good to use cemented tubs (ideally 6 to $8^{/}$ x $3^{/}$ x $2.5^{/}$), though the process can be done by heap and pit method also. Common among the organic wastes to be vermicompost are crop residues, kitchen wastes, market wastes, wastes of the agrobased industries, weeds, poultry wastes, livestock wastes and cattle dung. Cow dung slurry is used at the start of the process. Worms are to be released @ 1 kg or 1000 number per ton of the residues after noticing its partial decomposition (usually 10 to 15 days after properly placing the chopped wastes). Water is to be sprinkled to maintain 70 to 80% moisture content, but to be stopped when nearly 95% of the residue is decomposed (usually 5 days before harvest). Maturity is judged visually by observing the formation of granular structure of the compost on the surface of the tub/heap/pit. Generally, it takes two and half to three months time for the process to be completed. The prepared vermicompost is collected by scrapping of the produce in layers starting from the top, sieved to remove the earthworms, cocoons and eggs from the compost, dried under shade and then bagged. One vermicompost unit with 3 beds or tubs does produce about two and half to three and half tons vermicompost depending upon the nature of the residues. According to 'Organic Fertilizer Standards of FCO, 1985 amended in July 2021', a vermicompost for sale in India must have 15 to 25% moisture, total organic carbon 18%, total N 1%, total P_2O_5 0.8% and total K_2O 0.8% (all per cent by weight, minimum). Often a byproduct of a vermicomposting unit is vermiwash. Vermiwash, a rich source of micronutrients, is a liquid collected by passing water through a column of worm action. Vermiwash is very useful as a foliar spray.

225. Vermiwash – Vermiwash is a honey brown colour worm coelomic liquid extract prepared from vermicomposts. The worm worked out soils of vermicomposting have burrows formed by the earthworms. The bacteria inhabit these burrows extensively and enrich them biologically, especially with plant nutrients. The water passing through these passages washes the soluble materials present in the burrows and is collected as vermiwash. Vermiwash is a rich source of nitrogen, phosphorus and micronutrients, plant growth regulators, enzymes, humic substances, and mucous secretions of earthworms that prove very beneficial for the development of favourable soil structure. Vermiwash is an ecofriendly liquid fertilizer and can be applied both as foliar spray and soil application.

226. Volcanic rocks – Volcanic rocks are igneous rocks formed by the cooling and condensation of magma over the earth's surface. Cooling is thus rapid, and the rocks are cracky and fine grained due to rapid crystallization. Basalt, andesite and rhyolite are important volcanic rocks. Volcanic rocks are known also as extrusive rocks.

227. Watershed – Watershed is defined as any surface area from which runoff from rainfall and all other sources is collected and drained to a common point. Watershed in essence is the land and water area which contributes runoff to a common point. All lands on earth are part of one watershed or the other. The terms drainage basin and catchment area are often used as synonyms for watershed. A watershed may be only a few hectares as in small ponds or hundreds of square kilometers as in rivers. Watersheds are classified depending upon the size such as Macro watershed (> 50,000 ha), Sub watershed (10,000 to 50,000 ha), Milli watershed (1000 to 10,000 ha), Micro watershed (100 to 1000 ha), and Mini watershed (upto 100 ha). Watersheds are classified depending upon their drainage, shape and pattern of use also.

228. Weatherability of Minerals – All minerals are not equally weatherable. For instance, quartz is a very weathering resistant mineral (in fact, one of the most weathering resistant minerals), while olivine is one of the least weathering resistant minerals. Obviously, there must be certain factors to make a mineral susceptible or resistant to weathering. Some of the factors influencing the weatherability of a mineral are the minerals' degree of tetrahedral linkage, basicity, presence of iron and other oxidizable elements, density of the crystal packing, silica-alumina ratio, order of crystallization and physical characteristics like particle size, hardness, density etc.

229. Weathering - Physical disintegration and chemical decomposition of rocks and minerals, present mostly at the earth's surface, caused by physical agents and chemical processes, is called 'weathering'.

Weathering is of three kinds viz. physical weathering (or, mechanical weathering), chemical weathering and biological weathering. Weathering precedes the formation of regolith from rocks and minerals enroute to the formation of soils.

230. Wetlands - Wetlands are the areas where water stagnates over the soil or where water is present at or near the soil surface all along the year or for varying periods of time during the year including during the crop growing season. Permanent or periodical water saturation or near water saturation of the wetlands changes the soil properties, and creates conditions that favour the growth of specially adapted plants, and creates an environment much different from the surrounding areas that is unique for specialized animal forms. Two general categories of wetlands are (i) coastal or tidal wetlands and (ii) inland or nontidal wetlands. Coastal wetlands are found along the coast line of the sea, ocean, gulf, and bay. Inland wetlands are common (i) on floodplains along rivers and streams (riparian wetlands), (ii) in isolated depressions surrounded by drylands, (iii) along the boundary lines of lakes and ponds, and (iv) in low lying areas where precipitation sufficiently saturates the soil. Wetlands provide specialized habitat for thousands of aquatic and terrestrial plants and animals, such as fish, reptiles, and birds including the migratory birds. Wetlands are important for flood protection, water quality improvement, shoreline erosion check, aesthetics and recreational needs. Wetlands include mangroves, marshes, swamps and bogs. According to the criteria of Ramsar Convention, India does have seventy five internationally recognized wetlands (Ramsar sites). Among these wetlands, the Sundarbans in West Bengal, Chilika in Odisha, Kolleru in Andhra Pradesh and Vembanad Kol in Kerala are particularly important.

231. Wilting point – Wilting point represents the soil moisture content at which plant roots can no longer extract water at a rate sufficient to meet their needs inclusive of transpiration needs, and the plants exhibit wilting symptoms. At wilting point, the soil water resides in small capillary pores and around the soil particles. The corresponding matric potential of the soil moisture content at wilting point is – 15 bar (approximately equal to a pF of 4.19).

232. Wind Erosion Equation - Wind Erosion Equation (WEE) furnishes an approximate estimate of soil loss by wind erosion. The equation based on the field works for decades on soil loss by wind erosion was formulated by Woodruff and Siddoway in 1965. The equation quantifies soil loss by wind erosion as a function of five variables. Thus,

$$E = f(I, K, C, L, V)$$

In the above expression, E = estimated soil loss in tons/acre/year, I = soil erodibility index based on factors like soil surface texture, $CaCO_3$ content, and clay content, K = ridge roughness factor that depends on ridge spacing, height and erosive wind directions in relation to the ridge direction, C = climatic factor, a measure of the erosive potential of the wind speed and surface moisture, L = unsheltered distance (length) across an erodible field, measured along the prevailing wind erosion direction, and V = vegetative cover factor accounting for the kind, amount and orientation of growing plants and plant residues on the soil surface. Normally, wind and soil characteristics *C* and *I* are beyond one's control. The other factors however, can be controlled through the choice of cultural practices.

233. Windbreak – A windbreak is essentially a linear row of trees and shrubs intended for protection from the wind. Windbreaks retard the velocity of wind and thus lower its impact by providing a mechanical obstruction. Windbreaks thus reduce soil erosion, protect livestock from heat and cold, shield buildings and roads, provide habitat for wildlife, beautify the landscape by adding to aesthetics, conserve soil water, create carbon bank, provide travel routes and serve as source of wood and food. In the design of a windbreak, the location, orientation to the wind, height, width, density, and selection of species (of trees and shrubs) are some of the important aspects to be considered keeping in mind the purpose for which the windbreak is to be primarily used.

234. Zeta Potential – Zeta Potential, known also as the electrokinetic potential, is the electrical potential in the interfacial double layer of a colloidal system. Zeta potential is thus the potential difference between the dispersion medium and the stationery layer of the fluid attached to the particle layer. Zeta potential is the key factor in processes such as the preparation of colloidal dispersions, utilization of colloidal phenomena and the destruction of unwanted colloidal dispersions. Knowledge on zeta potential of colloidal systems does have considerable practical application value also. Thus, zeta potential analysis helps establishing optimum coagulation conditions for the removal of particulate matter and organic dyestuff from aqueous waste products such as industrial effluents. Knowledge on zeta potential is therefore of much use in water purification e. g. waste water treatment plants. Typically, higher the zeta potential value, the more stable the colloidal system is. When the zeta potential is zero, the dispersed phase of the colloidal system precipitates in the form of a solid. Zeta potential is measured in terms of milli volts (mV) and is represented by the Greek letter ζ.

VI. Expressions for Deriving Numerical Values of Soil Characteristics

Furnish the formulae for the numerical calculation of following soil characteristics

1. Bulk density - $\text{B. D.} = \frac{100\ \text{P. D.} - \%\ \text{pore space x P.D.}}{100}$

2. Particle density - $\text{P. D.} = \frac{100\ \text{B. D.}}{100 - \%\ \text{pore space}}$

3. Porosity - $\text{Porosity} = 1 - \frac{\text{B.D.}}{\text{P.D.}}$

4. Per cent Pore space- $\%\ \text{Pore space} = 100\ (1 - \frac{\text{B.D.}}{\text{P.D.}})$

5. Per cent Base Saturation –

$$\text{PBS} = (\frac{\text{Exchangeable } [Ca^{2+}] + [Mg^{2+}] + [K^+] + [Na^+]}{\text{Cation exchange capacity}})\ \text{x}\ 100$$

(concentrations of all the five are expressed in m. e./100 g soil or cmol (p^+)/kg soil)

6. Exchangeable Sodium Per centage - $\text{ESP} = \frac{\text{Exchangeable } [Na^+]}{\text{Cation exchange capacity}}\ \text{x}\ 100$

(concentrations of both exchangeable Na^+ and CEC are expressed in m. e./100 g soil or cmol (p^+)/kg soil)

7. Sodium Adsorption Ratio - $\text{SAR} = \frac{[Na^+]}{\sqrt{\frac{[Ca^{2+}] + [Mg^{2+}]}{2}}}$

(concentrations of exchangeable Na^+, Ca^{2+} and Mg^{2+} are expressed in m. e./100 g soil or cmol (p^+)/kg soil or per L water as the case may be)

8. Soluble Sodium Per centage of irrigation water -

$$\text{SSP} = \frac{[Na^+]}{[Ca^{2+}] + [Mg^{2+}] + [Na^+] + [K^+]}\ \text{x}\ 100$$

(provided (i) other cations are present in negligible amounts and (ii) the concentrations of all the four cations are expressed in m. e. per L)

9. Potassium Adsorption Ratio - $KAR = \frac{[K^+]}{\sqrt{\frac{[Ca^{2+}]+[Mg^{2+}]}{2}}}$

(concentrations of all the three cations are expressed in m. e./100 g soil)

10. Residual Sodium Carbonate -

$$RSC = ([CO_3^{2-}] + [HCO_3^-]) - ([Ca^{2+}] + [Mg^{2+}])$$

(all the four ionic constituents are expressed in m. e./100 g soil or cmol (p^+/e^-)/kg soil or m.e per L water as the case may be)

11. Residual Sodium Bicarbonate - $RSBC = [HCO_3^-] - [Ca^{2+}]$

(both the ionic constituents are expressed in m. e./100 g soil or m.e cmol (p^+/e^-)/kg soil or m.e per L water as the case may be)

12. Coefficient of Linear Extensibility - $COLE = [(\frac{L_m - L_d)}{L_d})] \times 100$

(L_m = length of a bar shaped soil when moistened to its plastic limit, and L_d = length of the bar shaped soil when it is air dried).

13. Stokes' law - $v = \frac{2\,(D_p - D_f)\,gr^2}{9\,\eta}$, where

(v = terminal velocity of the falling particle in the fluid, D_p = density of the particle, D_f = density of the fluid, g = acceleration due to gravity, r = radius of the falling spherical particle and η = viscosity of the fluid).

14. Darcy's law - $Q = c\,K\,A\,\frac{h_2 - h_1}{L} = \frac{c\,K\,A\,\Delta P}{L}$

(Q = flow velocity, c = a dimensionless proportionality constant, K = hydraulic conductivity, A = cross sectional area of the porous medium, h_1 and h_2 = heights above a datum of water = hydraulic gradient = Δ h = the pressure gradient = ΔP, and L = length over which the hydraulic drop or the pressure drop is taking place.

15. Fick's law of diffusion - $N = -D\frac{dc}{dx}$ or, $N = -D\frac{\Delta c}{\Delta x}$

(N is the molar flux of the gas normal to the x-direction, *dc/dx* = the concentration gradient, and D = the proportionality constant or diffusion coefficient.

16. Leaching Requirement for reclamation of saline soils -

$$LR = \frac{D_{dw}}{D_{iw}} = \frac{EC_{iw}}{EC_{dw}}$$

(D_{dw} = depth of the drainage water, D_{iw} = depth of the irrigation water, EC_{iw} = electrical conductivity of the irrigation water and EC_{dw} = electrical conductivity of the drainage water)

For very generalized calculations of LR for the reclamation of saline soils, EC_{dw} is taken as two times of the maximum permissible value of EC of saturation extract of the soil that can be tolerated by the crop in the field upto the point of no acceptable yield reduction. Therefore, if a field crop, say barley, (i) can tolerate the maximum EC of 9 mmhos/cm of a saturated extract of the soil, and (ii) the EC of the applied irrigation water is 2 mmhos/cm, the LR is 11.1 per cent. Stated otherwise, at least 11.1% of the applied irrigation water is to be leached down.

17. Permeability Index (P. I.) of irrigation water -

$$P.\ I. = \frac{[Na^+] + \sqrt{[HCO_3^-]}}{[Na^+] + [Ca^{2+}] + [Mg^{2+}]} \times 100$$

(the concentrations of the ions in terms of m.e. per L)

VII. Chemical Reactions Involved in Various Processes

Furnish the chemical reactions involved in the manufacture of the following fertilizers/ determination of the following soil chemical characteristics

1 Manufacture of ammonium sulphate by Gypsum process

(i) Formation of ammonium carbonate

$2\ NH_3 + CO_2 + H_2O \rightarrow (NH_4)_2CO_3$

(ii) Reaction of ammonium carbonate with gypsum to form ammonium sulphate

$(NH_4)_2CO_3 + CaSO_4 \rightarrow (NH_4)_2SO_4 + CaCO_3$

2. Manufacture of urea from NH_3 and CO_2

(i) Formation of ammonium carbamate

$2\ NH_3 + CO_2 \rightarrow H_2NCOONH_4$

(ii) Dehydration of ammonium carbamate to form urea

$H_2NCOONH_4 \rightarrow CO(NH_2)_2 + H_2O$

3. Manufacture of Sulphate of Potash from langbeinite ($K_2SO_4.2\ MgSO_4$) and KCl

$K_2SO_4.2\ MgSO_4 + 4\ KCl \rightarrow 3\ K_2SO_4 + 2\ MgCl_2$

4. Manufacture of single superphosphate [$Ca(H_2PO_4)_2 + CaSO_4$] from rock phosphate [$Ca_{10}(PO_4)_6F_2$]

Acidulation of rock phosphate with H_2SO_4

$Ca_{10}(PO_4)_6F_2 + 7\ H_2SO_4 \rightarrow 3\ Ca(H_2PO_4)_2 + 7\ CaSO_4 + 2\ HF$

or,

$Ca_3(PO_4)_2 + 2\ H_2SO_4 \rightarrow Ca(H_2PO_4)_2 + 2\ CaSO_4$

5. Manufacture of triple super phosphate [$Ca(H_2PO_4)_2$] from rock phosphate [$Ca_{10}(PO_4)_6F_2$]

Acidulation of rock phosphate with H_3PO_4

$Ca_{10}(PO_4)_6F_2 + 14\ H_3PO_4 \rightarrow 10\ Ca(H_2PO_4)_2 + 2\ HF$

6. Manufacture of nitrophosphates from rock phosphate [$Ca_{10}(PO_4)_6F_2$] by ODDA process

(i) Acidulation of rock phosphate with HNO_3

$$Ca_{10}(PO_4)_6F_2 + 14\ HNO_3 \rightarrow 3\ Ca(H_2PO_4)_2 + 7\ Ca(NO_3)_2 + 2\ HF$$

(ii) Removal of a part of $Ca(NO_3)_2$ from the reaction medium by crystallization process

(iii) Ammoniation of $Ca(H_2PO_4)_2$ and $Ca(NO_3)_2$ to form dicalcium phosphate ($CaHPO_4$) and NH_4NO_3

$$Ca(H_2PO_4)_2 + Ca(NO_3)_2 + 2\ NH_3 \rightarrow 2\ CaHPO_4 + 2\ NH_4NO_3$$

7. Manufacture of nitrophosphates from rock phosphate [$Ca_{10}(PO_4)_6F_2$] by PEC process (multi acid process)

(i) Acidulation of rock phosphate with HNO_3-H_2SO_4 mixture (3:1 mole ratio)

$$Ca_{10}(PO_4)_6F_2 + 12\ HNO_3 + 4H_2SO_4 \rightarrow 6\ H_3PO_4 + 4\ CaSO_4 + 6\ Ca(NO_3)_2 + 2\ HF$$

(ii) Ammoniation of the contents to form dicalcium phosphate ($CaHPO_4$), monoammonium phosphate and NH_4NO_3

$$6\ H_3PO_4 + 6\ Ca(NO_3)_2 + 2\ HF + 4\ CaSO_4 + 13\ NH_3 \rightarrow 5\ CaHPO_4 + NH_4H_2PO_4 + 12\ NH_4NO_3 + CaF_2 + 4\ CaSO_4$$

8. Determination of the total nitrogen content of soils/manure/plant samples inclusive of nitrate nitrogen by Kjeldahl Method

Four reactions are involved viz. (i) digestion (oxidation) of organic-N to $(NH_4)_2SO_4$ by conc. H_2SO_4, (ii) liberation of NH_3 from $(NH_4)_2SO_4$ by adding an alkali, (iii) formation of $NH_4H_2BO_3$ from NH_3 by adding boric acid, and (iv) titration of $NH_4H_2BO_3$ against standard H_2SO_4. Inclusion of nitrate N involves four additional reactions viz. (v) formation of nitrosalicylic acid, (vi) formation of H_2SO_3 (from H_2SO_4 and $Na_2S_2O_3$), (vii) formation of aminosalicylic acid, and (viii) formation of $(NH_4)_2SO_4$ from aminosalicylic acid.

(i) Organic nitrogen + conc. $H_2SO_4 \rightarrow (NH_4)_2SO_4$

(ii) $(NH_4)_2SO_4 + 2\ NaOH = 2\ NH_3\uparrow + Na_2SO_4 + 2\ H_2O$

(iii) $NH_3 + H_3BO_3 = NH_4H_2BO_3$

(iv) $2\ NH_4H_2BO_3 + H_2SO_4 = (NH_4)_2SO_4 + 2\ H_3BO_3$

(v) $C_6H_4OH.COOH + HNO_3 = C_6H_3OH.NO_2.COOH + H_2O$

(vi) $Na_2S_2O_3 + H_2SO_4 = Na_2SO_4 + H_2SO_3 + S$

(vii) $C_6H_3OH.NO_2.COOH + 3\ H_2SO_3 + H_2O = C_6H_3OH.NH_2.COOH + 3\ H_2SO_4$

(viii) $2\ C_6H_3OH.NH_2.COOH + 27\ H_2SO_4 = (NH_4)_2SO_4 + 14\ CO_2 + 26\ SO_2 + 30\ H_2O$

Factor: 1 ml 1 N H_2SO_4 = 0.014 g N

9. Determination of organic carbon content of soils by Walkley and Black's Method

Two reactions are involved viz. (i) oxidation of organic C (shown as C) to CO_2 by a known excess of $K_2Cr_2O_7$ and (ii) finding out the amount of $K_2Cr_2O_7$ consumed for the oxidation of organic C by titration of the unused $K_2Cr_2O_7$ against $FeSO_4$ (rather, ferrous ammonium sulphate).

(i) $3\ C + 2\ K_2Cr_2O_7 + 8\ H_2SO_4 = 2\ Cr_2(SO_4)_3 + 3\ CO_2 + 2\ K_2SO_4 + 8\ H_2O$

(ii) $K_2Cr_2O_7 + 6\ FeSO_4 + 7\ H_2SO_4 = Cr_2(SO_4)_3 + 3\ Fe_2(SO_4)_3 + K_2SO_4 + 7\ H_2O$

Factor: 1 ml 1 N $K_2Cr_2O_7$ = 0.003 g organic C = 0.005172 g organic matter

10. Determination of P_2O_5 content of soils/fertilizers (by volumetric method by using ammonium molybdate (Pemberton's method)

Three reactions are involved viz. (i) formation of a precipitate of ammonium phosphomolybdate, (ii) dissolution of the precipitate by a known excess of alkali, and (iii) finding out the amount of alkali consumed for dissolution of the precipitate by determining the amount of the unused alkali.

(i) $Ca(H_2PO_4)_2 + 24\ (NH_4)_2MoO_4 + 44\ HNO_3 = 2\ (NH_4)_3PO_4.12MoO_3\downarrow + 42\ NH_4NO_3 + Ca(NO_3)_2 + 24\ H_2O$

(ii) $(NH_4)_3PO_4.12MoO_3 + 23\ NaOH$ (known excess) $= 11\ Na_2MoO_4 + NaNH_4HPO_4 + (NH_4)_2MoO_4 + 11\ H_2O$ (+ unused NaOH)

(iii) NaOH (unused) + $HNO_3 = NaNO_3 + H_2O$

Factor: 1 ml 1 N NaOH = 0.0013467 g P = 0.0030839 g P_2O_5

11. Determination of CEC of soils by NH_4^+ saturation followed by NH_3 distillation method

The reactions involved are (i) saturation of the soil exchange complexes by NH_4^+ by replacing the adsorbed cations by exhaustive washing with ammonium acetate, (ii) replacement of the NH_4^+ adsorbed by K^+ by exhaustive washing with KCl, (iii) liberating NH_3 from NH_4^+ in the leachate solution by adding MgO and (iv) determining the liberated NH_3 by usual ammonia distillation method as in the case of the determination of total N content of soils.

(i) and (ii)

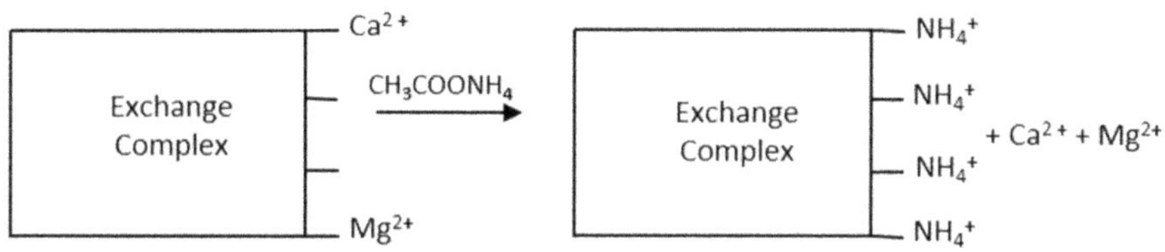

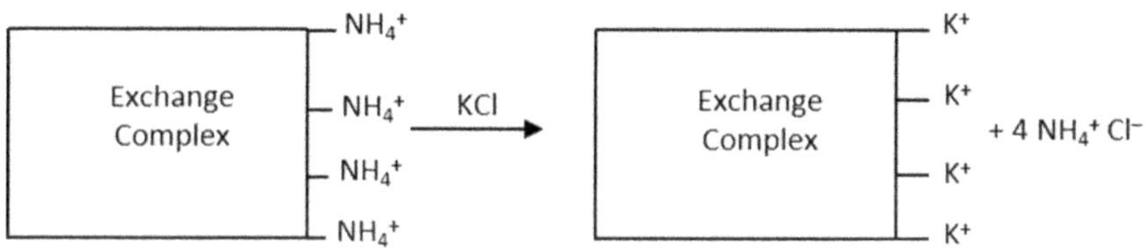

(iii) $NH_4Cl + OH^- \rightarrow NH_3\uparrow + H_2O + Cl^-$

(iv) and (v) – as in the case total N determination [reactions (iii) and (iv) of 8.

Factor: 1 ml 1 N H_2SO_4 = 1 milli equivalent = 0.1 cmole of +ve charges

IX. Appendix

1. Periodic Table of the Elements

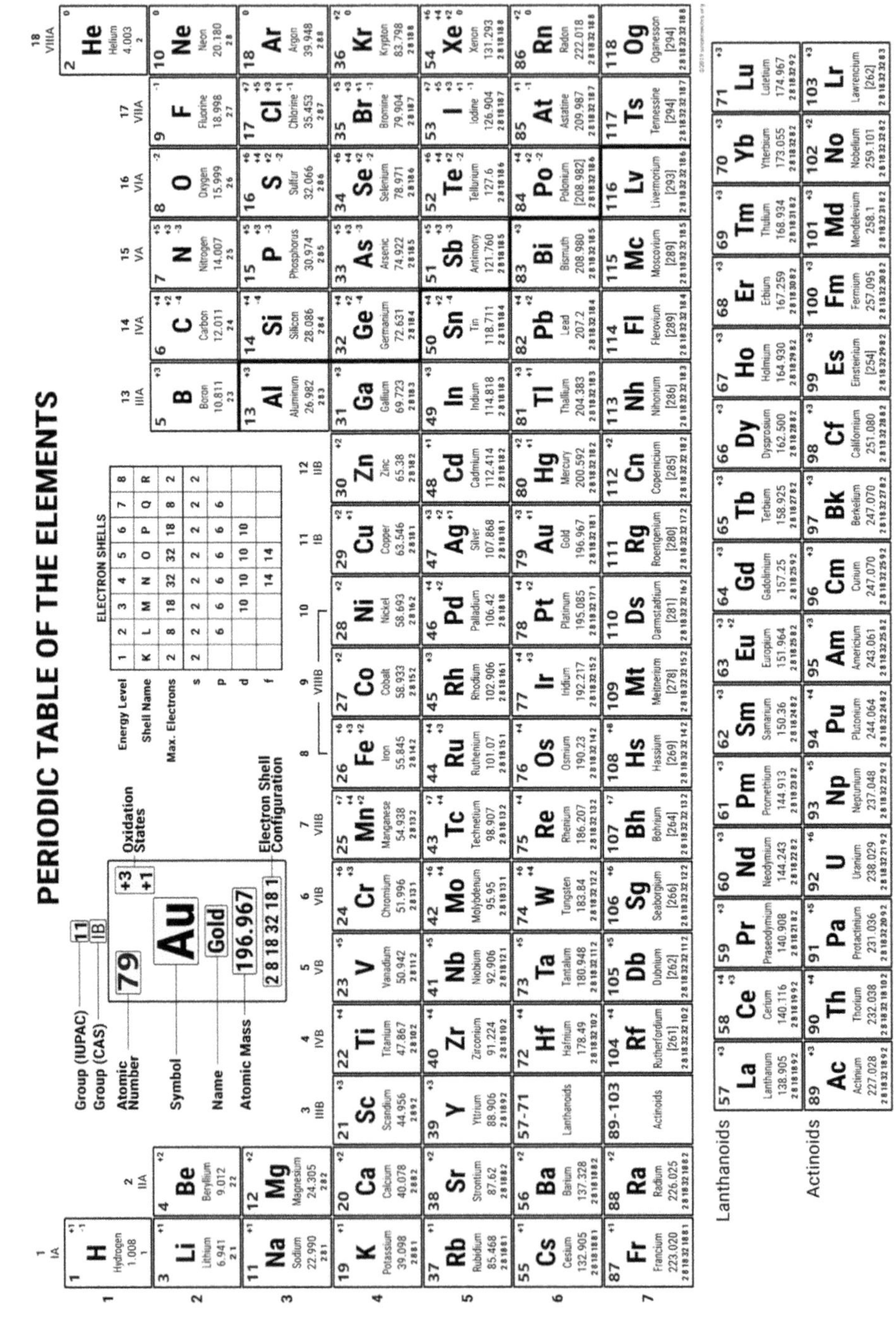

2. Atomic weights of the elements (in ascending atomic number)

Element	Symbol	Atomic No.	Atomic weight
Hydrogen	H	1	1.008
Helium	He	2	4.003
Lithium	Li	3	6.941
Beryllium	Be	4	9.012
Boron	B	5	10.811
Carbon	C	6	12.011
Nitrogen	N	7	14.007
Oxygen	O	8	15.999
Fluorine	F	9	18.998
Neon	Ne	10	20.18
Sodium	Na	11	22.99
Magnesium	Mg	12	24.305
Aluminum	Al	13	26.982
Silicon	Si	14	28.086
Phosphorus	P	15	30.974
Sulfur	S	16	32.066
Chlorine	Cl	17	35.453
Argon	Ar	18	39.948
Potassium	K	19	39.098
Calcium	Ca	20	40.078
Scandium	Sc	21	44.956
Titanium	Ti	22	47.867
Vanadium	V	23	50.942
Chromium	Cr	24	51.996
Manganese	Mn	25	54.938
Iron	Fe	26	55.845
Cobalt	Co	27	58.933
Nickel	Ni	28	58.693
Copper	Cu	29	63.546
Zinc	Zn	30	65.38
Gallium	Ga	31	69.723
Germanium	Ge	32	72.631
Arsenic	As	33	74.922
Selenium	Se	34	78.971
Bromine	Br	35	79.904
Krypton	Kr	36	83.798
Rubidium	Rb	37	85.468
Strontium	Sr	38	87.62
Yttrium	Y	39	88.906

Element	Symbol	Atomic No.	Atomic weight
Zirconium	Zr	40	91.224
Niobium	Nb	41	92.906
Molybdenum	Mo	42	95.95
Technetium	Tc	43	98.907
Ruthenium	Ru	44	101.07
Rhodium	Rh	45	102.906
Palladium	Pa	46	106.42
Silver	Ag	47	107.868
Cadmium	Cd	48	112.414
Indium	In	49	114.818
Tin	Sn	50	118.711
Antimony	Sb	51	121.76
Tellurium	Te	52	127.6
Iodine	I	53	126.904
Xenon	Xn	54	131.293
Cesium	Cs	55	132.905
Barium	Ba	56	137.328
Lanthanum	La	57	138.905
Cerium	Ce	58	140.116
Praseodymium	Pr	59	140.908
Neodymium	Nd	60	144.243
Promethium	Pm	61	144.913
Samarium	Sm	62	150.36
Europium	Eu	63	151.964
Gadolinium	Gd	64	157.25
Terbium	Tb	65	158.925
Dysprosium	Dy	66	162.5
Holmium	Ho	67	164.93
Erbium	Er	68	167.269
Thulium	Tm	69	168.934
Ytterbium	Yb	70	173.055
Lutetium	Lu	71	174.967
Hafnium	Hf	72	178.49
Tantalum	Ta	73	180.948
Tungsten	W	74	183.84
Rhenium	Re	75	186.207
Osmium	Os	76	190.23
Iridium	Ir	77	192.217
Platinum	Pt	78	195.085
Gold	Au	79	196.967
Mercury	Hg	80	200.592

Element	Symbol	Atomic No.	Atomic weight
Thallium	Tl	81	204.383
Lead	Pb	82	207.2
Bismuth	Bi	83	208.98
Polonium	Po	84	208.982
Astatine	At	85	209.987
Radon	Rn	86	220.018
Francium	Fr	87	223.02
Radium	Ra	88	226.025
Actinium	Ac	89	227.028
Thorium	Th	90	232.038
Protactinium	Pa	91	231.036
Uranium	U	92	238.029
Neptunium	Np	93	237.048
Plutonium	Pu	94	244.064
Americium	Am	95	243.061
Curium	Cm	96	247.07
Berkelium	Bk	97	247.07
Californium	Cf	98	251.08
Einsteinium	Es	99	252.083
Fermium	Fm	100	257.095
Mendelevium	Md	101	258.1
Nobelium	No	102	259.101
Lawrencium	Lr	103	266.119
Rutherfordium	Rf	104	[261]
Dubnium	Db	105	[262]
Seaborgium	Sg	106	[266]
Bohrium	Bh	107	[264]
Hassium	Hs	108	[269]
Meitnerium	Mt	109	[278]
Darmstadtium	Ds	110	281.162
Roengenium	Rg	111	281.168
Copernicium	Cn	112	285.174
Nihobium	Unt	113	286.181
Flerovium	Fl	114	[289]
Moscovium	Mc	115	[289]
Livermorium	Lv	116	[293]
Tennessine	Ts	117	[294]
Organesson	Og	118	[294]
Ununennium	Uu	119	[316]
Unbinilium	Ub	120	[320]
Unbiunium	Ub	121	[320]

Element	Symbol	Atomic No.	Atomic weight
Unbibium	Ub	122	[321]
Unbitrium	Ub	123	[325]
Unbiquadium	Ub	124	[330]
Unbipentium	Ub	125	[332]
Unbihexium	Ub	126	[334]
Unbiseptium	Ub	127	[335]

Atomic No. 57 to 71 - Lanthanides

Atomic No. 89 to 103 – Actinides

Figures in brackets indicate the mass number of the longest-lived isotope. Such elements do not have known stable isotopes.

3. Purity and specific gravity of commercial grades of concentrated acids and aqueous ammonia

Acid	Conc. % (m/m)	Specific gravity	Normality	Volume (ml) for 1 L 1 N solution
Acetic acid, CH_3COOH	99.7	1.05	17.43	57.5
Aqueous ammonia, NH_4OH	29.0	0.90	7.45	15.4
Hydrochloric acid, HCl	35.0	1.18	11.35	88.1
Hydrofluoric acid, HF	49.5	1.15	28.45	35.2
Nitric acid, HNO_3	70.5	1.42	15.89	63.0
Perchloric acid, $HClO_4$	71.0	1.67	11.89	84.1
Phosphoric acid, H_3PO_4	85.0	1.71	44.50	22.5
Sulphuric acid, H_2SO_4	96.5	1.84	36.21	27.8

4. Fundamental Physical Constants

Constant	Standard Symbol	Value
Avogadro's Number	N	6.02252 x 10^{23} mol^{-1}
Curie	Ci	3.70 x 10^{10} dps
Dalton	-	1.661 x 10^{-24} g
Faraday's constant	F	96,487 Coulomb mol^{-1}; 23,062 cal V^{-1} mol^{-1}
Universal Gas Constant	R	8.31434 J mol^{-1} $^{0}K^{-1}$; 1.98717 cal mol^{-1} $^{0}K^{-1}$
Planck's constant	h	6.62607 x 10^{-34} J sec; 1.62607 x 10^{-27} erg s 4.135667 x 10^{-15} eV s
Velocity of light (in vacuum)	c	2.99792 x 10^{8} m s^{-1}
Normal atmosphere, pressure	P	1,013,250 dyne cm^{-2}
Atomic Mass Unit	AMU/amu	1.6604 x 10^{-24} g
Electronic charge	e	4.80298 x 10^{-10}esu; 1.60210 x 10^{-19} coulomb
Electron rest mass	m_e	9.10908 x 10^{-31} kg
Proton rest mass	m_p	1.67252 x 10^{-27} kg
Neutron rest mass	m_n	1.67482 x 10^{-27} kg
Rydberg constant	$R\infty$	1.09737 x 10^{7} m^{-1}
Boltzmann constant	k	1.38054 x 10^{-23} J $^{0}K^{-1}$; 1.38054 x 10^{-16} erg $^{0}K^{-1}$
Acceleration of gravity	g	980.665 cm s^{-2}

5. Interconversions of Units of measurement

Conversion between	Multiply by* (unless mentioned)	Conversion between	Multiply by
Inch to centimeter Centimeter to inch	2.542 0.394	Meter to yard Yard to meter	1.094 0.914
Kilometer to mile Mile to kilometer	0.621 1.609	Angstrom to nanometer Nanometer to Angstrom	0.1 10
Hectare to acre Acre to hectare	0.405 2.469	Acre to square yard Acre to square meter	4847 4050
Square meter to square feet Square foot to square meter	10.760 0.093	Cubic meter to cubic feet Cubic feet to cubic meter	35.714 0.028
Cubic feet to liter Liter to cubic feet	28.3 0.035	Cubic inch to cubic centimeter Cubic centimeter to cubic inch	16.387 0.061
Gallon to liter Liter to gallon	3.78 0.265	Pound to gram Kilogram to pound	453.5 2.210
Ounce to gram Gram to ounce	28.4 0.035	Ton to Mega gram Mega gram to ton	0.907 1.102
Kg ha^{-1} to lbs $acre^{-1}$ Lbs $acre^{-1}$ to kg ha^{-1}	0.893 1.120	Liter ha^{-1} to gallon $acre^{-1}$ Gallon to liter ha^{-1}	0.107 9.346
Bar to pounds $inch^{-1}$ (psi) Bar to kPa	14.503 100	Bar to atmosphere Bar to mm of Hg	0.987 750.019
^0C to ^0K ^0K to ^0C	$^0C + 273^0$ $^0K - 273$	Joule to calorie Calorie to Joule	0.239 4.184
Joule to erg	0.735	Newton to dyne	10^5
Siemens m^{-1} to Millimhos cm^{-1} dS m^{-1} to millimhos cm^{-1}	10 1	ppm to mg kg^{-1} per cent to ppm	1 10^4
Becquerel to Curie Curie to Becquerel	2.7×10^{-11} 3.7×10^{10}	Gray to rad Rad to gray	100 0.01

6. Prefixes for Exponential Notations

Prefix	Abbreviation	Multiplied by	Prefix	Abbreviation	Multiplied by
deca	da	10^{1}	deci	d	10^{-1}
hecto	h	10^{2}	centi	c	10^{-2}
kilo	k	10^{3}	milli	m	10^{-3}
mega	M	10^{6}	micro	μ	10^{-6}
giga	G	10^{9}	nano	n	10^{-9}
tera	T	10^{12}	pico	p	10^{-12}
peta	P	10^{15}	femto	f	10^{-15}
exa	E	10^{18}	atto	a	10^{-18}
zetta	Z	10^{21}	zepto	z	10^{-21}
yotta	Y	10^{24}	yocto	y	10^{-24}

7. Greek alphabets

Alpha	α	Beta	β	Gamma	γ
Delta	δ	Epsilon	ε	Zeta	ζ
Eta	η	Theta	θ	Iota	ι
Kappa	κ	Lambda	λ	Mu	μ
Nu	ν	Xi	ξ	Omicron	ο
Pi	π	Rho	ρ	Sigma	σ
Tau	τ	Upsilon	υ	Phi	φ
Chi	χ	Psi	ψ	Omega	ω

www.ingramcontent.com/pod-product-compliance
Ingram Content Group UK Ltd.
Pitfield, Milton Keynes, MK11 3LW, UK
UKHW021321190726
13845UKWH00003B/15

9 789349 562004